T0012439

Birds *of* Louisiana & Mississippi

Field Guide

Stan Tekiela

Adventure Publications
Cambridge, Minnesota

Acknowledgments

Many thanks to the National Wildlife Refuge System along with state and local agencies, both public and private, for stewarding the lands that are critical to the many bird species we so love.

Edited by Sandy Livoti and Dan Downing

Cover, book design and illustrations by Jonathan Norberg

Range maps produced by Anthony Hertzel

Cover photo: Painted Bunting by Stan Tekiela

All photos by Stan Tekiela except p. 342 (juvenile) by **Rick & Nora Bowers**; pp. 194 (displaying), 238 (chick-feeding adult & juvenile) by **Dudley Edmondson**; p. 244 by **Frode Jacobsen/Shutterstock.com**; p. 46 (drying) by **JayPierstorff/Shutterstock.com**; pp. 46 (soaring), 260 (winter) by **Kevin T. Karlson**; pp. 156 (female), 300 (in flight) by **Brian E Kushner/Shutterstock.com**; pp. 72 (breeding), 168 (breeding) by **Ivan Kuzmin/Shutterstock.com**; p. 86 (in flight) by **Brian Lasenby/Shutterstock.com**; p. 170 (breeding) by **Brian E. Small**; p. 280 (displaying) by **Hartmut Walter**; pp. 46 (juvenile), 48 (juvenile), 178 (in-flight juvenile & juvenile), 294 (juvenile), 296 (in-flight juvenile) by **Brian K. Wheeler**; and pp. 224 (female), 284 (main), 360 (female) by **Jim Zipp**.

To the best of the publisher's knowledge, all photos were of live birds. Some were photographed in a controlled condition.

10 9 8 7 6 5 4 3 2 1

Birds of Louisiana & Mississippi Field Guide
First Edition 2011
Second Edition 2022
Copyright © 2011 and 2022 by Stan Tekiela
Published by Adventure Publications
An imprint of AdventureKEEN
310 Garfield Street South
Cambridge, Minnesota 55008
(800) 678-7006
www.adventurepublications.net
All rights reserved
Printed in China
ISBN 978-1-64755-299-2 (pbk.); ISBN 978-1-64755-300-5 (ebook)

TABLE OF CONTENTS

WHAT'S NEW?

It is hard to believe that it's been more than 10 years since the debut of *Birds of Louisiana & Mississippi Field Guide*. This critically acclaimed field guide has helped countless people identify and enjoy the birds that we love. Now, in this expanded second edition, *Birds of Louisiana & Mississippi Field Guide* has many new and exciting changes and a fresh look, while retaining the same familiar, easy-to-use format.

To help you identify even more birds in Louisiana and Mississippi, I have added 6 new species and more than 150 new color photographs. All of the range maps have been meticulously reviewed, and many updates have been made to reflect the ever-changing movements of the birds.

Everyone's favorite section, "Stan's Notes," has been expanded to include even more natural history information. "Compare" sections have been updated to help ensure that you correctly identify your bird, and additional feeder information has been added to help with bird feeding. I hope you will enjoy this great new edition as you continue to learn about and appreciate our Louisiana & Mississippi birds!

WHY WATCH BIRDS IN LOUISIANA AND MISSISSIPPI?

Millions of people have discovered bird feeding. It's a simple and enjoyable way to bring the beauty of birds closer to your home. Watching birds at your feeder and listening to them often leads to a lifetime pursuit of bird identification. The *Birds of Louisiana & Mississippi Field Guide* is for those who want to identify the common birds of Louisiana and Mississippi.

There are over 1,100 species of birds found in North America. In Louisiana there have been more than 450 different kinds of birds recorded throughout the years, and in Mississippi over 390 species have been reported. Both of these are an impressive number of species for a single state! These bird sightings were diligently recorded by hundreds of bird watchers throughout the state and became part of the official state records. From these valuable records, I have chosen 146 of the most common and easily seen birds of Louisiana and Mississippi to include in this field guide.

Bird watching, often called birding, is one of the most popular activities in America. Its outstanding appeal in Louisiana and Mississippi is due, in part, to an unusually rich and abundant birdlife. Why are there so many birds in these states? One reason is water—both salt water and fresh. The coastal areas of Louisiana and Mississippi attract larger varieties of birds and are home to many ocean-loving birds such as colony-nesting Royal Terns and surf-running Sanderlings. Along the coast are large stands of cypress forests surrounded by brackish wetlands where the White Ibis, Yellow-crowned Night-Heron and other species thrive. Grand Isle, a barrier island, is one of the greatest places to see birds migrating in spring. Hundreds of bird species stop here on their way north, making it one of the best places to see birds in the entire state.

In addition, Louisiana and Mississippi have many sizable lakes. Freshwater lakes in Louisiana cover more that 8,275 square miles (21,500 sq. km). In Mississippi, along with thousands of freshwater and saltwater marshes, there are four major lakes. These are man-made and cover approximately 290 square miles (754 sq. km). All of this water attracts millions of birds, such as Tricolored Herons and Belted Kingfishers. Each year tens of thousands of ducks and geese spend the winter in the Sabine, Mississippi, Atchafalaya, Ouachita, Big Black, Pearl and Red Rivers and the surrounding tributaries, making these places wonderful destinations to see birds during winter.

The forested region in northern and central parts of Louisiana and Mississippi is a good place to see woodland birds such as Pileated Woodpeckers and Eastern Phoebes. Driskill Mountain, the highest point in Louisiana, and Woodall Mountain, the highest point in Mississippi, are in this region. Look for Yellow-bellied Sapsuckers and White-breasted Nuthatches here.

Located between the forested region in northern Louisiana and Mississippi and the coast in the south are the coastal plains. Characterized by gently rolling hills, grasslands mark the transition zone between the north and the south. Grasslands dominating this part of Louisiana and Mississippi are home to birds such as Eastern Meadowlarks and Eastern Bluebirds.

Louisiana and Mississippi are one of the best places in North America to see a wide array of birds. Whether witnessing a nesting colony of herons and egrets along the coast or welcoming back wintering shorebirds and Ruby-throated Hummingbirds, bird watchers enjoy variety and excitement in Louisiana and Mississippi as each season turns to the next.

OBSERVE WITH A STRATEGY:
TIPS FOR IDENTIFYING BIRDS

Identifying birds isn't as difficult as you might think. By simply following a few basic strategies, you can increase your chances of successfully identifying most birds that you see. One of the first and easiest things to do when you see a new bird is to note **its color**. This field guide is organized by color, so simply turn to the right color section to find it.

Next, note the **size of the bird.** A strategy to quickly estimate size is to compare different birds. Pick a small, a medium and a large bird. Select an American Robin as the medium bird. Measured from bill tip to tail tip, a robin is 10 inches (25 cm). Now select two other birds, one smaller and one larger. Good choices are a House Sparrow, at about 6 inches (15 cm), and an American Crow, around 18 inches (45 cm). When you see a species you don't know, you can now quickly ask yourself, "Is it larger than a sparrow but smaller than a robin?" When you look in your field guide to identify your bird, you would check the species that are roughly 6–10 inches (15–25 cm). This will help to narrow your choices.

Next, note the **size, shape and color of the bill.** Is it long or short, thick or thin, pointed or blunt, curved or straight? Seed-eating birds, such as Northern Cardinals, have bills that are thick and strong enough to crack even the toughest seeds. Birds that sip nectar, such as Ruby-throated Hummingbirds, need long, thin bills to reach deep into flowers. Hawks and owls tear their prey with very sharp, curving bills. Sometimes, just noting the bill shape can help you decide whether the bird is a woodpecker, finch, grosbeak, blackbird or bird of prey.

Next, take a look around and note the **habitat** in which you see the bird. Is it wading in a marsh? Walking along a riverbank?

Soaring in the sky? Is it perched high in the trees or hopping along the forest floor? Because of diet and habitat preferences, you'll often see robins hopping on the ground but not usually eating seeds at a feeder. Or you'll see a Blue Grosbeak sitting on a tree branch but not climbing headfirst down the trunk, like a White-breasted Nuthatch would.

Noticing **what the bird is eating** will give you another clue to help you identify the species. Feeding is a big part of any bird's life. Fully one-third of all bird activity revolves around searching for food, catching prey and eating. While birds don't always follow all the rules of their diet, you can make some general assumptions. Northern Flickers, for instance, feed on ants and other insects, so you wouldn't expect to see them visiting a seed feeder. Other birds, such as Barn and Tree Swallows, eat flying insects and spend hours swooping and diving to catch a meal.

Sometimes you can identify a bird by **the way it perches.** Body posture can help you differentiate between an American Crow and a Red-tailed Hawk, for example. Crows lean forward over their feet on a branch, while hawks perch in a vertical position. Consider posture the next time you see an unidentified large bird in a tree.

Birds in flight are harder to identify, but noting the **wing size and shape** will help. Wing size is in direct proportion to body size, weight and type of flight. Wing shape determines whether the bird flies fast and with precision, or slowly and less precisely. Barn Swallows, for instance, have short, pointed wings that slice through the air, enabling swift, accurate flight. Turkey Vultures have long, broad wings for soaring on warm updrafts. House Finches have short, rounded wings, helping them to flit through thick tangles of branches.

Some bird species have a unique **pattern of flight** that can help in identification. American Goldfinches fly in a distinctive undulating pattern that makes it look like they're riding a roller coaster.

While it's not easy to make all of these observations in the short time you often have to watch a "mystery" bird, practicing these identification methods will greatly expand your birding skills. To further improve your skills, seek the guidance of a more experienced birder who can answer your questions on the spot.

BIRD BASICS

It's easier to identify birds and communicate about them if you know the names of the different parts of a bird. For instance, it's more effective to use the word "crest" to indicate the set of extra-long feathers on top of a Northern Cardinal's head than to try to describe it.

The following illustration points out the basic parts of a bird. Because it is a composite of many birds, it shouldn't be confused with any actual bird.

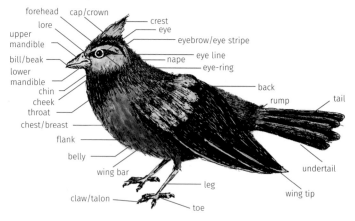

Bird Color Variables

No other animal has a color palette like a bird's. Brilliant blues, lemon yellows, showy reds and iridescent greens are common in the bird world. In general, male birds are more colorful than their female counterparts. This helps males attract a mate, essentially saying, "Hey, look at me!" Color calls attention to a male's health as well. The better the condition of his feathers, the better his food source, territory and potential for mating.

Male and female birds that don't look like each other are called sexually dimorphic, meaning "two forms." Dimorphic females often have a nondescript dull color, as seen in Indigo Buntings. Muted tones help females hide during the weeks of motionless incubation and draw less attention to them when they're out feeding or taking a break from the rigors of raising the young.

The males of some species, such as the Downy Woodpecker, Blue Jay and Bald Eagle, look nearly identical to the females. In woodpeckers, the sexes are differentiated by only a red mark, or sometimes a yellow mark. Depending on the species, the mark may be on top of the head, on the face or nape of neck, or just behind the bill.

During the first year, juvenile birds often look like their mothers. Since brightly colored feathers are used mainly for attracting a mate, young non-breeding males don't have a need for colorful plumage. It's not until the first spring molt (or several years later, depending on the species) that young males obtain their breeding colors.

Both breeding and winter plumages are the result of molting. Molting is the process of dropping old, worn feathers and replacing them with new ones. All birds molt, typically twice a year, with the spring molt usually occurring in late winter. At this time, most birds produce their brighter breeding plumage, which lasts throughout the summer.

Winter plumage is the result of the late summer molt, which serves a couple of important functions. First, it adds feathers for warmth in the coming winter season. Second, in some species it produces feathers that tend to be drab in color, which helps to camouflage the birds and hide them from predators. The winter plumage of the male American Goldfinch, for example, is olive-brown, unlike its canary-yellow breeding color during summer. Luckily for us, some birds, such as the male Northern Cardinal, retain their bright summer colors all year long.

Bird Nests

Bird nests are a true feat of engineering. Imagine constructing a home that's strong enough to weather storms, large enough to hold your entire family, insulated enough to shelter them from cold and heat, and waterproof enough to keep out rain. Think about building it without blueprints or directions and using mainly your feet. Birds do this!

Before building, birds must select an appropriate site. In some species, such as the House Wren, the male picks out several potential sites and assembles small twigs in each. The "extra" nests, called dummy nests, discourage other birds from using any nearby cavities for their nests. The male takes the female around and shows her the choices. After choosing her favorite, she finishes the construction.

In other species, such as the Orchard Oriole, the female selects the site and builds the nest, while the male offers an occasional suggestion. Each bird species has its own nest-building routine that is strictly followed.

As you can see in these illustrations, birds build a wide variety of nest types.

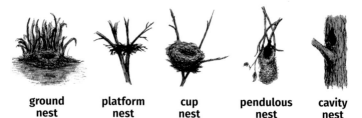

| ground nest | platform nest | cup nest | pendulous nest | cavity nest |

Nesting material often consists of natural items found in the immediate area. Most nests consist of plant fibers (such as bark from grapevines), sticks, mud, dried grass, feathers, fur, or soft, fuzzy tufts from thistle. Some birds, including Ruby-throated Hummingbirds, use spiderwebs to glue nest materials together.

Transportation of nesting material is limited to the amount a bird can hold or carry. Birds must make many trips afield to gather enough material to complete a nest. Most nests take four days or more, and hundreds, if not thousands, of trips to build.

A **ground nest** can be a mound of vegetation on the ground or in water. It can also be just a simple, shallow depression scraped out in earth, stones or sand. Killdeer and Black Skimmers scrape out ground nests without adding any nesting material.

The **platform nest** represents a much more complex type of construction. Typically built with twigs or sticks and branches, this nest forms a platform and has a depression in the center to nestle the eggs. Platform nests can be in trees; on balconies, cliffs, bridges, or man-made platforms; and even in flowerpots. They often provide space for the adventurous young and function as a landing platform for the parents.

Mourning Doves and herons don't anchor their platform nests to trees, so these can tumble from branches during high winds and storms. Hawks, eagles, Ospreys and other birds construct sturdier platform nests with large sticks and branches.

Other platform nests are constructed on the ground with mud, grass and other vegetation from the area. Many waterfowl build platform nests on the ground near or in water. A **floating platform nest** moves with the water level, preventing the nest, eggs and birds from being flooded.

Three-quarters of all songbirds construct a **cup nest,** which is a modified platform nest. The supporting platform is built first and attached firmly to a tree, shrub, or rock ledge or the ground. Next, the sides are constructed with grass, small twigs, bark or leaves, which are woven together and often glued with mud for added strength. The inner cup can be lined with down feathers, animal fur or hair, or soft plant materials and is contoured last.

The **pendulous nest** is an unusual nest that looks like a sock hanging from a branch. Attached to the end of small branches of trees, this unique nest is inaccessible to most predators and often waves wildly in a breeze.

Woven tightly with plant fibers, the pendulous nest is strong and watertight and takes up to a week to build. A small opening at the top or on the side allows parents access to the grass-lined interior. More commonly used by tropical birds, this complex nest has also been mastered by orioles and kinglets. It must be one heck of a ride to be inside one of these nests during a windy spring thunderstorm!

The **cavity nest** is used by many species of birds, most notably woodpeckers and Eastern Bluebirds. A cavity nest is often excavated from a branch or tree trunk and offers shelter from storms, sun, cold and predators. A small entrance hole in a tree can lead to a nest chamber that is up to a safe 10 inches (25 cm) deep.

Typically made by woodpeckers, cavity nests are usually used only once by the builder. Nest cavities can be used for many subsequent years by such inhabitants as Tree Swallows, mergansers and bluebirds. Kingfishers, on the other hand, can dig a tunnel up to 4 feet (1 m) long in a riverbank. The nest chamber at the end of the tunnel is already well insulated, so it's usually only sparsely lined.

One of the most clever of all nests is the **no nest,** or daycare nest. Parasitic birds, such as Brown-headed Cowbirds, don't build their own nests. Instead, the egg-laden female searches out the nest of another bird and sneaks in to lay an egg while the host mother isn't looking.

A mother cowbird wastes no energy building a nest only to have it raided by a predator. Laying her eggs in the nests of other birds transfers the responsibility of raising her young to the host. When she lays her eggs in several nests, the chances increase that at least one of her babies will live to maturity.

Who Builds the Nest?

Generally, the female bird constructs the nest. She gathers the materials and does the building, with an occasional visit from her mate to check on progress. In some species, both parents contribute equally to nest building. The male may forage for sticks, grass or mud, but it is the female that often fashions the nest. Only rarely does a male build a nest by himself.

Fledging

Fledging is the time between hatching and flight, or leaving the nest. Some species of birds are **precocial,** meaning they leave the nest within hours of hatching, though it may be weeks before they can fly. This is common in waterfowl and shorebirds.

Baby birds that hatch naked and blind need to stay in the nest for a few weeks (these birds are **altricial**). Baby birds that are still in the nest are **nestlings.** Until birds start to fly, they are called **fledglings.**

Why Birds Migrate

Why do so many species of birds migrate? The short answer is simple: food. Birds migrate to locations with abundant food, as it is easier to breed where there is food than where food is scarce. Summer Tanagers, for instance, are **complete migrators** that fly from the tropics of South America to nest in the forests of North America, where billions of newly hatched insects are available to feed to their young.

Other migrators, such as some birds of prey, migrate back to northern regions in spring. In these locations, they hunt mice, voles and other small rodents that are beginning to breed.

Complete migrators have a set time and pattern of migration. Every year at nearly the same time, they head to a specific wintering ground. Complete migrators may travel great distances, sometimes 15,000 miles (24,100 km) or more in one year.

Complete migration doesn't necessarily mean flying from Louisiana and Mississippi to a tropical destination. Dark-eyed Juncos, for example, are complete migrators that move from the far reaches of Canada to spend the winter here in Louisiana and Mississippi. This trip is still considered complete migration.

Complete migrators have many interesting aspects. In spring, males often leave a few weeks before the females, arriving early to scope out possibilities for nesting sites and food sources, and to begin to defend territories. The females arrive several weeks later. In many species, the females and their young leave earlier in the fall, often up to four weeks before the adult males.

Other species, such as the American Goldfinch, are **partial migrators**. These birds usually wait until their food supplies dwindle before flying south. Unlike complete migrators, partial migrators move only far enough south, or sometimes east and west, to find abundant food. In some years it might be only a few hundred miles, while in other years it can be as much as a thousand. This kind of migration, dependent on weather and the availability of food, is sometimes called seasonal movement.

Unlike the predictable complete migrators or partial migrators, **irruptive migrators** can move every third to fifth year or, in some cases, in consecutive years. These migrations are triggered when times are tough and food is scarce. Red-breasted Nuthatches are irruptive migrators. They leave their normal northern range in search of more food or in response to overpopulation.

Many other birds don't migrate at all. Carolina Chickadees, for example, are **non-migrators** that remain in their habitat all year long and just move around as necessary to find food.

How Do Birds Migrate?

One of the many secrets of migration is fat. While most people are fighting the ongoing battle of the bulge, birds intentionally gorge themselves to gain as much fat as possible without losing the ability to fly. Fat provides the greatest amount of energy per unit of weight. In the same way that your car needs gas, birds are propelled by fat and stall without it.

During long migratory flights, fat deposits are used up quickly, and birds need to stop to refuel. This is when backyard bird feeding stations and undeveloped, natural spaces around our towns and cities are especially important. Some birds require up to 2–3 days of constant feeding to build their fat reserves before continuing their seasonal trip.

Many birds, such as most eagles, hawks, Ospreys, falcons and vultures, migrate during the day. Larger birds can hold more body fat, go longer without eating and take longer to migrate. These birds glide along on rising columns of warm air, called thermals, that hold them aloft while they slowly make their way north or south. They generally rest at night and hunt early in the morning before the sun has a chance to warm the land and create good soaring conditions. Daytime migrators use a combination of landforms, rivers, and the rising and setting sun to guide them in the right direction.

The majority of small birds, called **passerines,** migrate at night. Studies show that some use the stars to navigate. Others use the setting sun, and still others, such as pigeons, use Earth's magnetic field to guide them north or south.

While flying at night may not seem like a good idea, it's actually safer. First, there are fewer avian predators hunting for birds at night. Second, night travel allows time during the day to find food in unfamiliar surroundings. Third, wind patterns at night tend to be flat, or laminar. Flat winds don't have the turbulence of daytime winds and can help push the smaller birds along.

HOW TO USE THIS GUIDE

To help you quickly and easily identify birds, this field guide is organized by color. Refer to the color key on the first page, note the color of the bird, and turn to that section. For example, the male Rose-breasted Grosbeak is black-and-white with a red patch on his chest. Because the bird is mostly black-and-white, it will be found in the black-and-white section.

Each color section is also arranged by size, generally with the smaller birds first. Sections may also incorporate the average size in a range, which in some cases reflects size differences between male and female birds. Flip through the pages in the color section to find the bird. If you already know the name of the bird, check the index for the page number.

In some species, the male and female are very different in color. In others, the breeding and winter plumage colors differ. These species will have an inset photograph with a page reference and will be found in two color sections.

You will find a variety of information in the bird description sections. To learn more, turn to the sample on pp. 22–23.

Range Maps

Range maps are included for each bird. Colored areas indicate where the bird is frequently found. The colors represent the presence of a species during a specific season, not the density, or amount, of birds in the area. Green is used for summer, blue for winter, red for year-round and yellow for migration.

While every effort has been made to depict accurate ranges, these are constantly in flux due to a variety of factors. Changing weather, habitat, species abundance and availability of vital resources, such as food and water, can affect the migration and movement of local populations, causing birds to be found in areas that are atypical for the species. So please use the maps as intended—as general guides only.

female
p. 137

male

Common Name

Range Map   *Scientific name*   **Color Indicator**

YEAR-ROUND
SUMMER
MIGRATION
WINTER

Size: measurement is from head to tip of tail; wingspan may be listed as well

Male: brief description of the male bird; may include breeding, winter or other plumages

Female: brief description of the female bird, which is sometimes different from the male

Juvenile: brief description of the juvenile bird, which often looks like the adult female

Nest: kind of nest the bird builds to raise its young; who builds it; number of broods per year

Eggs: number of eggs you might expect to see in a nest; color and marking

Incubation: average days the parents spend incubating the eggs; who does the incubation

Fledging: average days the young spend in the nest after hatching but before they leave the nest; who does the most "childcare" and feeding

Migration: type of migrator: complete (seasonal, consistent), partial (seasonal, destination varies), irruptive (unpredictable, depends on the food supply) or non-migrator

Food: what the bird eats most of the time (e.g., seeds, insects, fruit, nectar, small mammals, fish) and whether it typically comes to a bird feeder

Compare: notes about other birds that look similar and the pages on which they can be found; may include extra information to aid in identification

Stan's Notes: Interesting natural history information. This could be something to look or listen for or something to help positively identify the bird. Also includes remarkable features.

female
p. 163

male

Eastern Towhee

Pipilo erythrophthalmus

YEAR-ROUND
WINTER

Size: 7–8" (18–20 cm)

Male: Mostly black with rusty-brown sides and a white belly. Long black tail with a white tip. Short, stout, pointed bill and rich, red eyes. White wing patches flash in flight.

Female: similar to male but brown instead of black

Juvenile: light brown, a heavily streaked head, chest and belly, long dark tail with white tip

Nest: cup; female builds; 2 broods per year

Eggs: 3–4; creamy white with brown markings

Incubation: 12–13 days; female incubates

Fledging: 10–12 days; male and female feed the young

Migration: non-migrator to partial in Louisiana and Mississippi; moves around to find food

Food: insects, seeds, fruit; visits ground feeders

Compare: American Robin (p. 279) is slightly larger. The Gray Catbird (p. 275) lacks a black "hood" and rusty sides. Common Grackle (p. 33) lacks a white belly and has a long thin bill. Male Rose-breasted Grosbeak (p. 59) has a rosy patch in center of chest.

Stan's Notes: Named for its distinctive "tow-hee" call (given by both sexes) but known mostly for its other characteristic call, which sounds like "drink-your-tea!" Will hop backward with both feet (bilateral scratching), raking up leaf litter to locate insects and seeds. The female broods, but male does the most feeding of young. In some southern coastal states, some have red eyes and others have white eyes. Red-eyed variety seen in Louisiana and Mississippi.

female
p. 159

male

YEAR-ROUND

Brown-headed Cowbird
Molothrus ater

Size: 7½" (19 cm)

Male: Glossy black with a chocolate-brown head. Dark eyes. Pointed, sharp gray bill.

Female: dull brown with a pointed, sharp, gray bill

Juvenile: similar to female but with dull-gray plumage and a streaked chest

Nest: no nest; lays eggs in nests of other birds

Eggs: 5–7; white with brown markings

Incubation: 10–13 days; host bird incubates eggs

Fledging: 10–11 days; host birds feed the young

Migration: non-migrator in Louisiana and Mississippi

Food: insects, seeds; will come to seed feeders

Compare: The male Red-winged Blackbird (p. 31) is slightly larger with red-and-yellow patches on upper wings. Common Grackle (p. 33) has a long tail and lacks the brown head. European Starling (p. 29) has a shorter tail.

Stan's Notes: Cowbirds are members of the blackbird family. Known as brood parasites, Brown-headed Cowbirds are the only parasitic birds in Louisiana and Mississippi. Brood parasites lay their eggs in the nests of other birds, leaving the host birds to raise their young. Cowbirds are known to have laid their eggs in the nests of over 200 species of birds. While some birds reject cowbird eggs, most incubate them and raise the young, even to the exclusion of their own. Look for warblers and other birds feeding young birds twice their own size. Named "Cowbird" for its habit of following bison and cattle herds to feed on insects flushed up by the animals. Numbers may increase during winter, when migratory birds from the north join resident birds for the season.

winter

breeding

European Starling
Sturnus vulgaris

YEAR-ROUND

Size: 7½" (19 cm)

Male: Glittering, iridescent purplish black in spring and summer; duller and speckled with white in fall and winter. Long, pointed, yellow bill in spring; gray in fall. Pointed wings. Short tail.

Female: same as male

Juvenile: similar to adults, with grayish-brown plumage and a streaked chest

Nest: cavity; male and female line cavity; 2 broods per year

Eggs: 4–6; bluish with brown markings

Incubation: 12–14 days; female and male incubate

Fledging: 18–20 days; female and male feed the young

Migration: non-migrator

Food: insects, seeds, fruit; visits seed or suet feeders

Compare: The Common Grackle (p. 33) has a long tail. Male Brown-headed Cowbird (p. 27) has a brown head. Look for the shiny, dark feathers to help identify the European Starling.

Stan's Notes: A common songbird. Mimics the songs of up to 20 bird species and imitates sounds, including the human voice. Often displaces woodpeckers, chickadees and other cavity-nesting birds. Jaws are more powerful when opening than when closing, enabling the bird to pry open crevices to find insects. Often displaces woodpeckers, chickadees and other cavity-nesting birds. Gathers in the hundreds in fall and winter. Large families gather with blackbirds in the fall. Not a native bird; 100 starlings were introduced to New York City in 1890–91 from Europe. Bill changes color in spring and fall.

female
p. 175

male

Red-winged Blackbird
Agelaius phoeniceus

Size: 8½" (22 cm)

Male: Jet black with red-and-yellow patches (epaulets) on upper wings. Pointed black bill.

Female: heavily streaked brown with a pointed brown bill and white eyebrows

Juvenile: same as female

Nest: cup; female builds; 2–3 broods per year

Eggs: 3–4; bluish green with brown markings

Incubation: 10–12 days; female incubates

Fledging: 11–14 days; female and male feed the young

Migration: non-migrator in Louisiana and Mississippi

Food: seeds, insects; visits seed and suet feeders

Compare: The male Brown-headed Cowbird (p. 27) is smaller and glossier and has a brown head. The bold red-and-yellow epaulets distinguish the male Red-winged from other blackbirds.

Stan's Notes: One of the most widespread and numerous birds in Louisiana and Mississippi. In fall and winter, migrant and resident Red-wingeds gather in huge numbers (thousands) with other blackbirds to feed in agricultural fields, marshes, wetlands, and lakes and rivers. Flocks with as many as 10,000 birds have been reported. Males defend their territory by singing from the tops of surrounding vegetation. The male repeats his call from the top of a cattail while showing off his red-and-yellow shoulder patches. The female chooses a mate and often builds her nest over shallow water in a thick stand of cattails. The male can be aggressive when defending the nest. Feeds mostly on seeds in spring and fall, and insects throughout the summer.

Common Grackle
Quiscalus quiscula

YEAR-ROUND

Size: 11–13" (28–33 cm)

Male: Large, iridescent blackbird with bluish-black head and purplish-brown body. Long black tail. Long, thin bill and bright-golden eyes.

Female: similar to male but smaller and duller

Juvenile: similar to female

Nest: cup; female builds; 2 broods per year

Eggs: 4–5; greenish white with brown markings

Incubation: 13–14 days; female incubates

Fledging: 16–20 days; female and male feed the young

Migration: non-migrator in Louisiana and Mississippi; will move around to find food

Food: fruit, seeds, insects; will come to seed and suet feeders

Compare: Male Boat-tailed (p. 39) and Great-tailed (p. 41) Grackles are larger and have a much longer tail. The European Starling (p. 29) is much smaller with a speckled appearance, and a yellow bill during breeding season. The male Red-winged Blackbird (p. 31) has red-and-yellow wing markings (epaulets).

Stan's Notes: Usually nests in small colonies of up to 75 pairs but travels with other blackbird species in large flocks. Known to feed in farm fields. The common name is derived from the Latin word *gracula*, meaning "jackdaw," another species of bird. The male holds his tail in a deep V shape during flight. The flight pattern is usually level, as opposed to an undulating movement. Unlike most birds, it has larger muscles for opening its mouth than for closing it, enabling it to pry crevices apart to find hidden insects.

Common Gallinule

Gallinula galeata

YEAR-ROUND
SUMMER
MIGRATION

Size: 13–15" (33–38 cm)

Male: Nearly black overall with yellow-tipped red bill. Red forehead. Thin line of white along sides. Yellowish-green legs.

Female: same as male

Juvenile: same as adult, but brown with white throat and dirty-yellow legs

Nest: ground; female and male build; 1–2 broods

Eggs: 2–10; brown with dark markings

Incubation: 19–22 days; female and male incubate

Fledging: 40–50 days; female and male feed the young

Migration: partial to non-migrator in parts of Louisiana and Mississippi; to coastal southern states

Food: insects, snails, seeds, green leaves, fruit

Compare: American Coot (p. 37) is similar in size but lacks the distinctive yellow-tipped bill and red forehead of Common Gallinule. Purple Gallinule (p. 117) is similar in size but has an iridescent blue-and-green body.

Stan's Notes: Also known as Mud Hen or Pond Chicken. A nearly all-black duck-like bird often seen in freshwater marshes and lakes. Walks on floating vegetation or swims while hunting for insects. Females known to lay eggs in other gallinule nests in addition to their own. Builds its nest with cattails and bulrushes and sometimes takes an old nest in a low shrub. A cooperative breeder, having young of first brood help raise young of second. Young leave nest usually within a few hours after hatching, but they stay with the family for a couple months. Young ride on backs of adults.

American Coot
Fulica americana

WINTER

Size: 13–16" (33–40 cm)

Male: Gray-to-black waterbird. Duck-like white bill with a dark band near the tip and a small red patch near the eyes. Small white patch near base of tail. Green legs and feet. Red eyes.

Female: same as male

Juvenile: much paler than adults, with a gray bill

Nest: floating platform; female and male construct; 1 brood per year

Eggs: 9–12; pinkish buff with brown markings

Incubation: 21–25 days; female and male incubate

Fledging: 49–52 days; female and male feed young

Migration: complete, to Louisiana and Mississippi, other southern states, Mexico and Central America

Food: insects, aquatic plants

Compare: Smaller than most waterfowl, it is the only black, duck-like bird with a white bill.

Stan's Notes: Usually seen in large flocks on open water. Not a duck, as it has large lobed toes instead of webbed feet. An excellent diver and swimmer, bobbing its head as it swims. A favorite food of Bald Eagles. It is not often seen in flight, unless it's trying to escape from an eagle. To take off, it scrambles across the surface of the water, flapping its wings. Gives a unique series of creaks, groans and clicks. Anchors its floating platform nest to vegetation. Huge flocks with as many as 1,000 birds gather for migration. Migrates at night. The common name "Coot" comes from the Middle English word *coote*, which was used to describe various waterfowl. Also called Mud Hen.

female
p. 197

male

Boat-tailed Grackle

Quiscalus major

YEAR-ROUND

Size: 15–17" (38–43 cm), male
13–15" (33–38 cm), female

Male: Iridescent blue-black bird. Bright-yellow or dark eyes. Very long keel-shaped tail.

Female: brown version of male, lacks iridescence

Juvenile: similar to female

Nest: cup; female builds; 2 broods per year

Eggs: 2–4; pale greenish blue with brown marks

Incubation: 13–15 days; female incubates

Fledging: 12–15 days; female feeds the young

Migration: non-migrator; moves around to find food

Food: insects, berries, seeds, fish; visits feeders

Compare: Male Common Grackle (p. 33) lacks Boat-tailed's distinctive long tail. Fish Crow (p. 43) and American Crow (p. 45) are similar but have a very different shape. Look for an iridescent blue head and a very long tail.

Stan's Notes: A noisy bird of coastal saltwater and inland marshes, giving several harsh, high-pitched calls and several squeaks. Eats a wide variety of foods, from grains to fish. Sometimes seen picking insects off the backs of cattle. Will also visit bird feeders. Makes a cup nest with mud or cow dung and grass. Nests in small colonies. Most nesting occurs from February through July and occasionally again from October to December. Much less widespread in Louisiana and Mississippi than the Common Grackle and the Great-tailed Grackle. Boat-taileds on the Gulf Coast have dark eyes, while Atlantic Coast birds have bright-yellow eyes.

female
p. 199

male

Great-tailed Grackle
Quiscalus mexicanus

YEAR-ROUND

Size: 18" (45 cm), male
15" (38 cm), female

Male: Large all-black bird with iridescent purple sheen on the head and back. Exceptionally long tail. Bright-yellow eyes.

Female: much smaller than the male, brown bird with gray to brown belly, light-brown-to-white eyes, eyebrows, throat and upper chest.

Juvenile: similar to female

Nest: cup; female builds; 1–2 broods per year

Eggs: 3–5; greenish blue with brown markings

Incubation: 12–14 days; female incubates

Fledging: 21–23 days; female feeds the young

Migration: non-migrator to partial in Louisiana; moves around to find food

Food: insects, fruit, seeds; comes to seed feeders

Compare: Common Grackle (p. 33) is smaller, with a much shorter tail. Male Brown-headed Cowbird (p. 27) lacks the long tail and has a brown head. The male Boat-tailed Grackle (p. 39) is found along the coast.

Stan's Notes: This is our largest grackle. It was once considered a subspecies of the Boat-tailed Grackle, which occurs along the Gulf Coast and in the southern part of Louisiana and Mississippi. Prefers to nest close to water in an open habitat. A colony nester. Males do not participate in nest building, incubation or raising young. Males rarely fight; females squabble over nest sites and materials. Several females mate with one male.

41

Fish Crow

Corvus ossifragus

YEAR-ROUND

Size: 16" (40 cm)

Male: All-black bird appearing nearly identical to the American Crow, but with a longer tail and a smaller head and bill.

Female: same as male

Juvenile: same as adult

Nest: platform; female and male construct; 1 brood per year

Eggs: 4–5; blue or gray-green with brown marks

Incubation: 16–18 days; female and male incubate

Fledging: 21–24 days; female and male feed the young

Migration: non-migrator

Food: insects, carrion, mollusks, berries, seeds

Compare: American Crow (p. 45) is nearly identical, but it is larger and has a shorter tail and a larger head and bill. Fish Crow is most easily differentiated from American Crow by its higher-pitched call.

Stan's Notes: Essentially a bird of the coast and along major rivers, but can be found across Louisiana and Mississippi. Less common farther away from the coast. Not uncommon for it to break open mollusk shells by dropping them onto rocks from above. Very sociable and gregarious. Nests in small colonies, often building a stick nest in a palm tree. Forms small winter flocks of up to 100 birds, unlike the American Crow, which often forms winter flocks of several hundred. The best way to distinguish between the two crow species is by their remarkably different calls. Fish Crow has a high, nasal "cah."

in flight

American Crow
Corvus brachyrhynchos

YEAR-ROUND

Size: 18" (45 cm)

Male: All-black bird with black bill, legs and feet. Can have a purple sheen in direct sunlight.

Female: same as male

Juvenile: same as adult

Nest: platform; female builds; 1 brood per year

Eggs: 4–6; bluish to olive-green with brown marks

Incubation: 18 days; female incubates

Fledging: 28–35 days; female and male feed the young

Migration: non-migrator; moves around to find food

Food: fruit, insects, mammals, fish, carrion; will come to seed and suet feeders

Compare: Fish Crow (p. 43) is nearly identical, but it is smaller, has a longer tail and a smaller head and bill. American Crow is most easily differentiated from the Fish Crow by its lower-pitched call.

Stan's Notes: One of the most recognizable birds in Louisiana and Mississippi, found in all habitats. Imitates other birds and human voices. One of the smartest of all birds and very social, often entertaining itself by provoking chases with other birds. Eats roadkill but is rarely hit by vehicles. Can live as long as 20 years. Often reuses its nest every year if it's not taken over by a Great Horned Owl. Collects and stores bright, shiny objects in the nest. Unmated birds, known as helpers, help to raise the young. Extended families roost together at night, dispersing daily to hunt. Best differentiated from the Fish Crow by its call. Cannot soar on thermals; flaps constantly and glides downward.

drying

juvenile

soaring

Black Vulture
Coragyps atratus

YEAR-ROUND

Size: 25–28" (63–71 cm); up to 5¼' wingspan

Male: Black with dark-gray head and legs. Short tail. In flight, all black with light-gray wing tips.

Female: same as male

Juvenile: similar to adult

Nest: no nest, on a stump or on ground, or takes an abandoned nest; 1 brood per year

Eggs: 1–3; light green with dark markings

Incubation: 37–45 days; female and male incubate

Fledging: 75–80 days; female and male feed the young

Migration: non-migrator

Food: carrion; occasionally will capture small live mammals

Compare: Turkey Vulture (p. 49) is slightly larger, with a bright-red head. Turkey Vulture has two-toned wings with a black leading edge and light-gray trailing edge. The Black Vulture has shorter, gray-tipped wings and a shorter tail.

Stan's Notes: Also called Black Buzzard. A more gregarious bird than the Turkey Vulture. In flight, the Black Vulture holds its wings straight out to its sides unlike the Turkey Vulture, which holds its wings in a V pattern. More aggressive while feeding but less skilled at finding carrion than the Turkey Vulture, it is thought the Black Vulture's sense of smell is less developed. Families stay together up to a year. Often nests and roosts with other Black Vultures. If startled, especially at the nest, it regurgitates with power and accuracy. Like other vultures, it often spreads its wings to warm up in the morning or dry out after a rain (see inset).

soaring

juvenile

drying

YEAR-ROUND

Turkey Vulture
Cathartes aura

Size: 26–32" (66–80 cm); up to 6' wingspan

Male: Large and black with a naked red head and legs. In flight, wings are two-toned with a black leading edge and a gray trailing edge. Wing tips end in finger-like projections. Tail is long and squared. Ivory bill.

Female: same as male but slightly smaller

Juvenile: similar to adults, with a gray-to-blackish head and bill

Nest: no nest or minimal nest, on a cliff or in a cave, sometimes in a hollow tree; 1 brood per year

Eggs: 1–3; white with brown markings

Incubation: 38–41 days; female and male incubate

Fledging: 66–88 days; female and male feed the young

Migration: non-migrator in Louisiana and Mississippi

Food: carrion; parents regurgitate to feed the young

Compare: Black Vulture (p. 47) has shorter wings and tail. Bald Eagle (p. 95) is larger and lacks two-toned wings. Look for the obvious naked red head to identify the Turkey Vulture.

Stan's Notes: The naked head reduces the risk of feather fouling (picking up diseases) from contact with carcasses. It has a strong bill for tearing apart flesh. Unlike hawks and eagles, it has weak feet more suited for walking than grasping. One of the few birds with a good sense of smell. Mostly mute, making only grunts and groans. Holds its wings in an upright V shape in flight. Teeters from wing tip to wing tip as it soars and hovers. Seen in trees with wings outstretched, sunning, drying after a rain (see inset). Numbers increase in winter, when resident birds are joined by northern migrants.

in flight

juvenile

crests

drying

Double-crested Cormorant
Phalacrocorax auritus

WINTER

Size: 31–35" (79–89 cm); up to 4⅓' wingspan

Male: Large black waterbird with unusual blue eyes and a long, snake-like neck. Large gray bill, with yellow at the base and a hooked tip.

Female: same as male

Juvenile: lighter brown with a grayish chest and neck

Nest: platform, in colony; male and female construct; 1 brood per year

Eggs: 3–4; bluish white without markings

Incubation: 25–29 days; female and male incubate

Fledging: 37–42 days; male and female feed the young

Migration: complete, to Louisiana and Mississippi

Food: small fish, aquatic insects

Compare: Male Anhinga (p. 53) has white spots and streaks and a long straight bill without a hooked tip. The Turkey Vulture (p. 49) is similar in size and also perches on branches with wings open to dry in sun, but it has a naked red head. American Coot (p. 37) lacks the long neck and long pointed bill.

Stan's Notes: Flocks fly in a large V or a line. Usually roosts in large colonies in trees close to water. Swims underwater to catch fish, holding its wings at its sides. This bird's outer feathers soak up water, but its body feathers don't. To dry off, it strikes an upright pose with wings outstretched, facing the sun (see inset). Gives grunts, pops and groans. Named "Double-crested" for the crests on its head, which are not often seen. "Cormorant" is a contraction from *corvus marinus*, meaning "crow" or "raven," and "of the sea."

51

male

female

juvenile

Anhinga

Anhinga anhinga

YEAR-ROUND
SUMMER

Size: 33–37" (84–94 cm); up to 3¾' wingspan

Male: All black with glossy green and white spots and streaks on shoulders and wings. Long neck and tail. Long, narrow yellow bill.

Female: similar to male, buff-brown neck and breast

Juvenile: similar to female, light-brown-to-white body

Nest: platform; female and male build; 1 brood per year

Eggs: 2–4; light blue without markings

Incubation: 26–29 days; female and male incubate

Fledging: 21–25 days; female and male feed the young

Migration: partial to non-migrator in Louisiana and Mississippi

Food: fish, aquatic insects, crustaceans and small mammals

Compare: The Double-crested Cormorant (p. 51) is slightly smaller and lacks the white spots and streaks of male Anhinga. Cormorant has a shorter bill with a curved tip unlike the long straight bill of the Anhinga.

Stan's Notes: Also called Snakebird due to its habit of appearing like a snake—surfacing with just its head and long thin neck showing above the water. It skewers fish, a favorite prey, with its long sharp bill. Unlike ducks and other diving birds, its feathers become waterlogged, which helps it when diving and maneuvering underwater. Afterward, it often strikes a pose with wings spread to dry in the sun (see photo). A strong flier and frequently seen soaring, it is confused with birds of prey. In flight, the long neck and tail help to identify it.

male

female

Black-and-white Warbler
Mniotilta varia

SUMMER
MIGRATION

Size: 5" (13 cm)

Male: Small with zebra-like striping and a black-and-white striped crown. Black cheek patch and chin. White belly.

Female: duller than male and lacks a black cheek patch and chin

Juvenile: similar to female

Nest: cup; female builds; 1 brood per year

Eggs: 4–5; white with brown markings

Incubation: 10–11 days; female incubates

Fledging: 9–12 days; female and male feed the young

Migration: complete, to southern states, Mexico and Central and South America

Food: insects

Compare: Climbs down tree trunks headfirst, like the Brown-headed Nuthatch (p. 245) and White-breasted Nuthatch (p. 249). Look for a small black-and-white bird climbing down trees to identify the Black-and-white Warbler.

Stan's Notes: The only warbler that moves headfirst down a tree trunk. Look for it searching for insect eggs in the bark of large trees. Its song sounds like a slowly turning, squeaky wheel. The female will perform a distraction dance to draw predators away from the nest. Constructs its nest on the ground, concealed beneath dead leaves or at the base of a tree. Most common during spring and fall migrations, but is also a common summer resident, nesting in northern Louisiana and Mississippi. Found in a variety of habitats.

male

female

YEAR-ROUND

Downy Woodpecker

Dryobates pubescens

Size: 6" (15 cm)

Male: Small woodpecker with a white belly and black-and-white spotted wings. Red mark on the back of the head and a white stripe down the back. Short black bill.

Female: same as male but lacks the red mark

Juvenile: same as female, some with a red mark near the forehead

Nest: cavity with a round entrance hole; male and female excavate; 1 brood per year

Eggs: 3–5; white without markings

Incubation: 11–12 days; female incubates during the day, male incubates at night

Fledging: 20–25 days; male and female feed the young

Migration: non-migrator

Food: insects, seeds; visits seed and suet feeders

Compare: The Hairy Woodpecker (p. 63) is larger. Look for the Downy's shorter, thinner bill.

Stan's Notes: Abundant and widespread where trees are present. This is perhaps the most common woodpecker in the U.S. Stiff tail feathers help to brace it like a tripod as it clings to a tree. Like other woodpeckers, it has a long, barbed tongue to pull insects from tiny places. Mates drum on branches or hollow logs to announce territory, which is rarely larger than 5 acres (2 ha). Repeats a high-pitched "peek-peek" call. Nest cavity is wider at the bottom than at the top and is lined with fallen wood chips. Male performs most of the brooding. During winter, it will roost in a cavity. Undulates in flight. A very common feeder bird in Louisiana and Mississippi.

female
p. 165

male

Rose-breasted Grosbeak

Pheucticus ludovicianus

MIGRATION

Size: 7–8" (18–20 cm)

Male: Plump black-and-white bird with a large triangular, rose-colored patch on the breast. Wing linings are rose-red. Large ivory bill.

Female: heavily streaked with obvious white eyebrows and orange-to-yellow wing linings

Juvenile: similar to female

Nest: cup; female and male construct; 1–2 broods per year

Eggs: 3–5; blue-green with brown markings

Incubation: 13–14 days; female and male incubate

Fledging: 9–12 days; female and male feed the young

Migration: complete, to Mexico, Central America and South America

Food: insects, seeds, fruit; comes to seed feeders

Compare: Male is very distinctive, with no look-alikes. Look for the rose breast patch to identify.

Stan's Notes: Seen in small groups across Louisiana and Mississippi during spring and fall migrations. Both sexes sing a rich, robin-like song with a chip note in the tune, but male is much louder and clearer than female. Often prefers to nest in mature deciduous forests. Males arrive first in small groups, females join them several days later. Several males will visit feeders at the same time in spring. Louisiana and Mississippi are important stopovers, so feeders should be kept full during this time. The rose breast patch varies in size and shape in each male. Male has white wing patches that flash in flight. "Grosbeak" refers to its thick, strong bill, used to crush seeds. Does not breed in Louisiana and Mississippi; nests in northern states.

male

female

WINTER

Yellow-bellied Sapsucker
Sphyrapicus varius

Size: 8–9" (20–23 cm)

Male: Checkered back with a red forehead, crown and chin. Yellow to tan on the chest and belly. White wing patches are seen flashing in flight.

Female: similar to male but with a white chin

Juvenile: similar to female, dull brown and lacks any red marking

Nest: cavity; female and male excavate, often in a live tree; 1 brood per year

Eggs: 5–6; white without markings

Incubation: 12–13 days; female incubates during the day, male incubates at night

Fledging: 25–29 days; female and male feed the young

Migration: complete, to Louisiana, Mississippi, other southern states, Mexico and Central America

Food: insects, tree sap; comes to suet feeders

Compare: The Red-headed Woodpecker (p. 65) has an all-red head. Look for the red chin and crown to identify the male Sapsucker, and the white chin and red crown to identify the female.

Stan's Notes: Found in small woods, forests, and suburban and rural areas. Drills rows of holes in trees to bleed the sap. Oozing sap attracts bugs, which it also eats. Defends its sapping sites from other birds that try to drink from the taps. Does not suck sap; rather, it laps the sticky liquid with its long, bristly tongue. A quiet bird, it makes few vocalizations but will meow like a cat. Drums on hollow branches, but unlike other woodpeckers, its rhythm is irregular. Makes short undulating flights with rapid wingbeats.

male

female

Hairy Woodpecker
Leuconotopicus villosus

YEAR-ROUND

Size: 9" (23 cm)

Male: Black-and-white woodpecker with a white belly. Black wings with rows of white spots. White stripe down the back. Long black bill. Red mark on the back of the head.

Female: same as male but lacks the red mark

Juvenile: grayer version of the female

Nest: cavity with an oval entrance hole; female and male excavate; 1 brood per year

Eggs: 3–6; white without markings

Incubation: 11–15 days; female incubates during the day, male incubates at night

Fledging: 28–30 days; male and female feed the young

Migration: non-migrator

Food: insects, nuts, seeds; comes to seed and suet feeders

Compare: Downy Woodpecker (p. 57) is much smaller and has a much shorter bill. Look for Hairy Woodpecker's long bill, nearly equal to the width of its head.

Stan's Notes: A common bird in wooded backyards. Announces its arrival with a sharp chirp before landing on feeders. Responsible for eating many destructive forest insects. Uses its barbed tongue to extract insects from trees. Tiny, bristle-like feathers at the base of the bill protect the nostrils from wood dust. Drums on hollow logs, branches or stovepipes in spring to announce territory. Often prefers to excavates nests cavities in live trees. Excavates a larger, more-oval-shaped entrance than the round hole of the Downy Woodpecker. Makes short flights from tree to tree.

juvenile

Red-headed Woodpecker
Melanerpes erythrocephalus

YEAR-ROUND

Size: 9" (22.5 cm)

Male: All-red head with a solid black back. White chest, belly and rump. Black wings with large white wing patches seen flashing in flight. Black tail. Gray legs and bill.

Female: same as male

Juvenile: gray brown with white chest, lacks any red

Nest: cavity; male builds with help from female; 1 brood per year

Eggs: 4–5; white without markings

Incubation: 12–13 days; female and male incubate

Fledging: 27–30 days; female and male feed the young

Migration: partial to non-migrator; will move to areas with an abundant supply of nuts

Food: insects, nuts, fruit; visits suet and seed feeders

Compare: No other woodpecker in Louisiana and Mississippi has an all-red head. The Pileated Woodpecker (p. 89) is the only other woodpecker with a solid black back, but it has a partially red head.

Stan's Notes: One of the few non-dimorphic woodpeckers, with males and females that look alike. Bill is strong enough to excavate a nest cavity only in soft, dead trees. Prefers open woodlands or edges with many dead branches. Nests later than its close relative, the Red-bellied Woodpecker, and will often take its cavity, if vacant. Unlike other woodpeckers, which use nest cavities just once, it may use the same cavity for several years in a row. Often perches on top of dead snags. Stores acorns and other nuts. Gives a shrill, hoarse "churr" call. Decreasing populations nationwide.

male

female

Red-bellied Woodpecker
Melanerpes carolinus

YEAR-ROUND

Size: 9–9½" (23–24 cm)

Male: Black-and-white "zebra-backed" woodpecker with a white rump. Red crown extends down the nape of the neck. Tan chest. Pale-red tinge on the belly, often hard to see.

Female: same as male but with a light-gray crown

Juvenile: gray version of adults; lacks a red crown and red nape

Nest: cavity; female and male excavate; 1 brood per year

Eggs: 4–5; white without markings

Incubation: 12–14 days; female incubates during the day, male incubates at night

Fledging: 24–27 days; female and male feed the young

Migration: non-migrator; moves around to find food

Food: insects, nuts, fruit; visits suet and seed feeders

Compare: Similar to the Northern Flicker (p. 187) and Yellow-bellied Sapsucker (p. 61). The Red-headed Woodpecker (p. 65) has an all-red head. Look for the zebra-striped back to help identify the Red-bellied Woodpecker.

Stan's Notes: Likes shady woodlands, forest edges and backyards. Digs holes in rotten wood to find spiders, centipedes, beetles and more. Hammers acorns and berries into crevices of trees for winter food. Returns to the same tree to excavate a new nest below that of the previous year. Undulating flight with rapid wingbeats. Gives a loud "querrr" call and a low "chug-chug-chug." Named for the pale red tinge on its belly. Expanding its range all over the country.

female

male

Scissor-tailed Flycatcher
Tyrannus forficatus

SUMMER
MIGRATION

Size: 9½" (24 cm)

Male: White-to-gray head, neck, breast and back. Black wings with bright pink wing linings, seen in flight. Faint pink coloring on flanks and belly. An extremely long black tail with patches of white.

Female: similar to the male, with much shorter tail

Juvenile: similar to adults; with shorter tail, lacking pink underwings and sides

Nest: cup; female builds; 1 brood per year

Eggs: 3–5; white with brown and red markings

Incubation: 14–17 days; female incubates

Fledging: 14–16 days; female and male feed the young

Migration: complete, to Mexico and Central America

Food: insects

Compare: This flycatcher's extremely long tail and the distinctive black-and-white pattern with its pink wing linings make it hard to confuse with any other bird.

Stan's Notes: A wonderful summer nester in Louisiana. Like most flycatchers, it hunts for insects by waiting on a post or low tree and flying out to capture them as the pass by. Drops to the ground to hunt for insects much more than other flycatchers. Male performs an up-down and zigzag courtship flight, showing off his long tail. Sometimes will end the flight with a reverse somersault. When not breeding, often seen in large flocks. Roosts communally, with up to 200 individuals. Closely related to kingbirds.

winter

breeding

Ruddy Turnstone

Arenaria interpres

MIGRATION
WINTER

Size: 9½" (24 cm)

Male: Breeding male has a white breast and belly with a black bib. Wings and back are black and chestnut. Head has a black-and-white marking. Orange legs. Slightly upturned black bill. Winter male has a brown-and-white head and breast pattern.

Female: similar to male but duller

Juvenile: similar to adults, but black-and-white head has a scaly appearance

Nest: ground; female builds; 1 brood per year

Eggs: 3–4; olive-green with dark markings

Incubation: 22–24 days; male and female incubate

Fledging: 19–21 days; male feeds the young

Migration: complete, to Louisiana and Mississippi, other southern coastal states, Mexico

Food: aquatic insects, fish, mollusks, crustaceans, worms, eggs

Compare: Unusually ornamented shorebird. Look for the striking black-and-white pattern on the head and neck, and orange legs to identify.

Stan's Notes: A common migrant and winter resident. Also known as Rock Plover. Named "Turnstone" because it turns stones over on rocky beaches to find food. Hangs around crabbing operations to eat scraps from nets. Can be tolerant of humans when feeding. Females often leave before their young leave the nest (fledge), resulting in males raising the young. Males have a bare spot on the belly (brood patch) to warm the young, something only females normally have.

winter
p. 283

breeding

Black-bellied Plover
Pluvialis squatarola

MIGRATION
WINTER

Size: 11–12" (28–30 cm)

Male: Striking black-and-white breeding plumage. Belly, breast, sides, face and neck are black. Cap, nape of neck, and belly near tail are white. Black legs and bill.

Female: less black on belly and breast than male

Juvenile: grayer than adults, with much less black

Nest: ground; male and female construct; 1 brood per year

Eggs: 3–4; pink or green with black-brown markings

Incubation: 26–27 days; male incubates during the day, female incubates at night

Fledging: 35–45 days; male feeds the young, the young learn quickly to feed themselves

Migration: complete, to Louisiana and Mississippi, the East and Gulf Coasts, West Indies

Food: insects

Compare: Breeding Dunlin (p. 171) is slightly smaller, with a rusty back and long down-curved bill. Look for Black-bellied Plover's large black patch on the belly, face and chest, and a white cap.

Stan's Notes: Males perform a "butterfly" courtship flight to attract females. Breeds at age 3. Female leaves male and young about 12 days after the eggs hatch. A winter resident along the Louisiana coast, dipping into Mississippi. Begins arriving in July and August (fall migration) and leaves in April. During flight, in any plumage, displays a white rump and stripe on wings with black axillaries (armpits). Can be very common on the beach during winter.

female
p. 201

male

Bufflehead
Bucephala albeola

WINTER

Size: 13–15" (33–38 cm)

Male: A small, striking duck with white sides and a black back. Greenish-purple head, iridescent in bright sun, with a large white head patch.

Female: brownish-gray with a dark brown head and white cheek patch behind the eyes

Juvenile: similar to female

Nest: cavity; female lines old woodpecker cavity; 1 brood per year

Eggs: 8–10; ivory to olive without markings

Incubation: 29–31 days; female incubates

Fledging: 50–55 days; female leads young to food

Migration: complete to Louisiana, Mississippi, other southern states and Mexico

Food: aquatic insects, crustaceans, mollusks

Compare: Male Hooded Merganser (p. 83) is larger and has rust-brown sides. Look for the large white bonnet-like patch on a greenish-purple head to help identify the male Bufflehead.

Stan's Notes: A small, common diving duck, almost always seen in small groups or with other duck species on rivers, ponds and lakes. A diving duck, seen during migration and winter. Nests in vacant woodpecker holes. When cavities in trees are scarce, known to use a burrow in an earthen bank or will use a nest box. Lines the cavity with fluffy down feathers. Unlike other ducks, the young stay in the nest for up to two days before they venture out with their mothers. The female is very territorial and remains with the same mate for many years. Does not breed in Louisiana and Mississippi.

Black-necked Stilt
Himantopus mexicanus

YEAR-ROUND
SUMMER
MIGRATION

Size: 14" (36 cm)

Male: Black-and-white with ridiculously long red-to-pink legs. Upper parts of the head, neck and back are black. Lower parts are white. Long black bill.

Female: similar to male but browner on back

Juvenile: similar to female but brown instead of black

Nest: ground; female and male construct; 1 brood per year

Eggs: 3–5; off-white with dark markings

Incubation: 22–26 days; male incubates during the day, female incubates at night

Fledging: 28–32 days; female and male feed the young

Migration: non-migrator to complete in Louisiana and Mississippi; to southern states, Mexico, and Central and South America

Food: aquatic insects

Compare: Outrageous length of the red-to-pink legs makes this shorebird hard to confuse with any other.

Stan's Notes: A breeding bird in Louisiana and Mississippi. Found along the East Coast and as far north as the Great Lakes. Nests alone or in small colonies in open areas. This very vocal bird of shallow marshes gives a "kek-kek-kek" call. Its legs are up to 10 inches (25 cm) long and may be the longest legs in the bird world in proportion to the body. Known to transport water with water-soaked belly feathers (belly-soaking) to cool eggs in hot weather. Aggressively defends its nest, eggs and young. Young leave the nest shortly after hatching.

female
p. 209

male

Lesser Scaup
Aythya affinis

WINTER

Size: 16–17" (40–43 cm)

Male: Appears mostly black with bold white sides and gray back. Chest and head look nearly black, but head appears purple with green highlights in direct sun. Bright-yellow eyes.

Female: overall brown with dull-white patch at base of light-gray bill, yellow eyes

Juvenile: same as female

Nest: ground; female builds; 1 brood per year

Eggs: 8–14; olive-buff without markings

Incubation: 22–28 days; female incubates

Fledging: 45–50 days; female teaches the young to feed

Migration: complete, to Louisiana and Mississippi, other southern states, Mexico, Central America

Food: aquatic plants and insects

Compare: The male Ring-necked Duck (p. 81) has a bold white ring around its bill, a black back and lacks bold white sides. The male Blue-winged Teal (p. 207) has a bright-white crescent-shaped mark at the base of bill.

Stan's Notes: Often in large flocks numbering in the thousands on lakes, ponds and sewage lagoons during winter. Prefers fresh water, but can be seen along the coast. Completely submerges itself to feed on the bottom of lakes (unlike dabbling ducks, which only tip forward to reach bottom). Note the bold white stripe under wings when in flight. Male leaves female when she starts incubating eggs. Quantity of eggs (clutch size) increases with age of female. Interesting baby-sitting arrangement in which groups of young (crèches) are tended by 1–3 adult females.

female
p. 213

male

Ring-necked Duck
Aythya collaris

WINTER

Size:	16–19" (41–48 cm)
Male:	Striking black duck with light-gray-to-white sides. Blue bill with a bold white ring and a thinner ring at the base. Peaked head with a sloped forehead.
Female:	brown with darker-brown back and crown, light-brown sides, gray face, white line behind eyes, white eye-ring, white ring around the bill, and peaked head
Juvenile:	similar to female
Nest:	ground; female builds; 1 brood per year
Eggs:	8–10; olive-gray to brown without markings
Incubation:	26–27 days; female incubates
Fledging:	49–56 days; female teaches the young to feed
Migration:	complete, to Louisiana and Mississippi, other southern states, Mexico and Central America
Food:	aquatic plants and insects
Compare:	Male Lesser Scaup (p. 79) is similar in size, but it has a gray back, unlike the black back of the male Ring-necked Duck. Look for the blue bill with a bold white ring to identify the male Ring-necked Duck.

Stan's Notes: One of the most abundant ducks in Louisiana and Mississippi during winter. Usually in larger freshwater lakes rather than saltwater marshes, in small flocks or just pairs. Watch for this diving duck to dive underwater to forage for food. Springs up off the water to take flight. Flattens its crown when diving. Named "Ring-necked" for its cinnamon collar, which is nearly impossible to see. Also called Ring-billed Duck due to the white ring on its bill.

female
p. 211

male

WINTER

Hooded Merganser
Lophodytes cucullatus

Size: 16–19" (40–48 cm)

Male: Black and white with rust-brown sides. Crest "hood" raises to show a large white patch on each side of the head. Long, thin, black bill.

Female: brown and rust with ragged, rust-red "hair" and a long, thin, brown bill

Juvenile: similar to female

Nest: cavity; female lines an old woodpecker cavity or a nest box near water; 1 brood per year

Eggs: 10–12; white without markings

Incubation: 32–33 days; female incubates

Fledging: 71 days; female feeds the young

Migration: complete, to Louisiana, Mississippi, and other southern states

Food: small fish, aquatic insects, crayfish

Compare: Male Bufflehead (p. 75) is smaller than the Hooded Merganser and has white sides. The male Wood Duck (p. 315) has a green head. The white patch on the head and rust-brown sides distinguish the male Hoodie.

Stan's Notes: A small diving bird of shallow ponds, sloughs, lakes and rivers, usually in small groups. The male voluntarily raises and lowers its crest to show off the large white head patch (hood). Quick, low flight across the water, with fast wingbeats. Male has a deep, rolling call. Female gives a hoarse quack. Nests in wooded areas. Female will lay some eggs in the nests of other Hooded Mergansers or Wood Ducks, resulting in 20–25 eggs in some nests. Rarely, she shares a nest, sitting with a Wood Duck. When young hatch, there is often a mixture of Hooded Mergansers and Wood Ducks.

skimming

YEAR-ROUND

Black Skimmer
Rynchops niger

Size: 18" (45 cm); up to 3½' wingspan

Male: A striking black-and-white bird with black on top and white on bottom. Very distinct black-tipped red bill with lower bill longer than the upper. Red legs tuck up and out of sight when in flight.

Female: similar to male but smaller

Juvenile: similar to adults, spotty brown on top

Nest: ground; female and male construct; 1 brood per year

Eggs: 3–5; bluish white with brown markings

Incubation: 21–23 days; female and male incubate

Fledging: 23–25 days; female and male feed the young

Migration: non-migrator; moves around to find food

Food: small fish, shrimp

Compare: Royal Tern (p. 345) has a similar shape, but it lacks the black back and black-tipped red bill. Look for a large black-and-white bird, with a lower bill longer than the upper bill, that skims across the water.

Stan's Notes: Also called Scissorbill or Razorbill, referring to this bird's unusual long bill. Uses its unique bill while in flight to cut through the water to catch fish or shrimp close to the surface. Commonly feeds with several other skimmers. Often seen flying to and from nesting colony with fish in its bill. Nests in large colonies, often associated with terns. Found along the Gulf and East Coasts.

in flight

YEAR-ROUND

American Oystercatcher
Haematopus palliatus

Size: 18–19" (45–48 cm)

Male: Large shorebird with dark-brown sides, wings and back, and a white chest and belly. Black head. Red ring around the eyes and a large red-orange bill. Pink legs.

Female: same as male

Juvenile: more gray than black and lacks the brightly colored bill

Nest: ground; male and female construct; 1 brood per year

Eggs: 2–4; olive with sparse brown markings

Incubation: 24–29 days; female incubates during the day, male incubates at night

Fledging: 35–40 days; male and female feed the young, the young learn quickly to feed themselves

Migration: non-migrator; moves around to find food

Food: shellfish, insects, aquatic insects, worms

Compare: Winter Black-bellied Plover (p. 283) is smaller. Look for the Oystercatcher's large and obvious red-orange bill.

Stan's Notes: A large, stunningly beautiful bird that stands out on the beach. This chunky shorebird has a flattened, heavy bill, which it uses to pry open shellfish and probe sand for insects and worms. Oystercatchers open oysters in two different ways. Some, known as "stabbers," sneak up on mollusks and stab their bills between shells before they have a chance to close. Others, called "hammerers," shatter one half of the shell.

male

female

Pileated Woodpecker

Dryocopus pileatus

YEAR-ROUND

Size: 19" (48 cm)

Male: Crow-size woodpecker with a black back and bright-red forehead, crest and mustache. Long gray bill. White leading edge of wings flashes brightly during flight.

Female: same as male but with a black forehead; lacks a red mustache

Juvenile: similar to adults but duller and browner

Nest: cavity; male and female excavate; 1 brood per year

Eggs: 3–5; white without markings

Incubation: 15–18 days; female incubates during the day, male incubates at night

Fledging: 26–28 days; female and male feed the young

Migration: non-migrator; moves around to find food

Food: insects; will come to suet and peanut feeders

Compare: The Red-headed Woodpecker (p. 65) is about half the size and has an all-red head. Look for the bright-red crest and exceptionally large size to identify the Pileated Woodpecker.

Stan's Notes: Our largest woodpecker. The common name comes from the Latin *pileatus*, which means "wearing a cap." A relatively shy bird that prefers large tracts of woodland. Drums on hollow branches, chimneys and so forth to announce its territory. Excavates oval holes up to several feet long in tree trunks, looking for insects to eat. Large wood chips lie on the ground by excavated trees. Favorite food is carpenter ants. Feeds regurgitated insects to its young. Young emerge from the nest looking just like the adults.

in flight

juvenile

YEAR-ROUND
SUMMER

Black-crowned Night-Heron

Nycticorax nycticorax

Size: 22–27" (56–69 cm); up to 3½' wingspan

Male: A stocky, hunched and inactive heron with black back and crown, white belly and gray wings. Long dark bill and bright-red eyes. Short dull-yellow legs. Breeding adult has 2 long white plumes on crown.

Female: same as male

Juvenile: golden-brown head and back with white spots; streaked breast; yellow-orange eyes and brown bill

Nest: platform; female and male build; 1 brood

Eggs: 3–5; light blue without markings

Incubation: 24–26 days; female and male incubate

Fledging: 42–48 days; female and male feed the young

Migration: non-migrator to partial in Louisiana and Mississippi

Food: fish, aquatic insects

Compare: Yellow-crowned Night-Heron (p. 301) has a similar size, a bold white cheek patch and lacks the black back of the Black-crowned. A perching Great Blue Heron (p. 305) looks twice the size of a Black-crowned. Look for a short-necked heron with a black back and crown.

Stan's Notes: A very secretive bird, this heron is most active near dawn and dusk (crepuscular). It hunts alone, but it nests in small colonies. Roosts in trees during the day. Often squawks if disturbed from the daytime roost. Often seen being harassed by other herons during the day. Stalks quiet backwaters in search of small fish and crabs.

soaring

Osprey
Pandion haliaetus

YEAR-ROUND
SUMMER
MIGRATION

Size: 21–24" (53–61 cm); up to 5½' wingspan

Male: Large eagle-like bird with a white chest, belly and head. Dark eye line. Nearly black back. Black "wrist" marks on the wings. Dark bill.

Female: same as male but slightly larger and with a necklace of brown streaks

Juvenile: similar to adults, with a light-tan breast

Nest: platform on a raised wooden platform, man-made tower or tall dead tree; female and male build; 1 brood per year

Eggs: 2–4; white with brown markings

Incubation: 32–42 days; female and male incubate

Fledging: 48–58 days; male and female feed the young

Migration: complete to non-migrator in Louisiana and Mississippi

Food: fish

Compare: The juvenile Bald Eagle (p. 95) is brown with white speckles. The adult Bald Eagle is on average 10 inches larger and has an all-white head and tail. Look for the white belly and dark eye line to identify the Osprey.

Stan's Notes: The only species in its family, and the only raptor that plunges into water feet-first to catch fish. Always near water. Can hover for a few seconds before diving. Carries fish in a head-first position for better aerodynamics. Wings angle back in flight. Often harassed by Bald Eagles for its catch. Gives a high-pitched, whistle-like call, often calling in flight as a warning. Mates have a long-term pair bond. Resident birds are joined by northern migrators in winter. Was nearly extinct but is now doing well.

soaring

juvenile

soaring
juvenile

Bald Eagle

Haliaeetus leucocephalus

YEAR-ROUND
WINTER

Size: 31–37" (79–94 cm); up to 7½' wingspan

Male: White head and tail contrast sharply with the dark-brown-to-black body and wings. Large, curved yellow bill and yellow feet.

Female: same as male but larger

Juvenile: dark brown with white speckles and spots on the body and wings; gray bill

Nest: massive platform, usually in a tree; female and male build; 1 brood per year

Eggs: 2–3; off-white without markings

Incubation: 34–36 days; female and male incubate

Fledging: 75–90 days; female and male feed the young

Migration: partial to non-migrator in Louisiana and Mississippi

Food: fish, carrion, birds (mainly ducks)

Compare: Black Vulture (p. 47) is smaller, has a shorter tail and lacks the adult Bald Eagle's white head and tail. Turkey Vulture (p. 49) is smaller and flies with its two-toned wings held in a V shape, unlike the straight-out wing position of the Bald Eagle.

Stan's Notes: Nearly became extinct due to DDT poisoning and illegal killing. Returns to the same nest each year, adding more sticks and enlarging it to huge proportions, at times up to 1,000 pounds (450 kg). In their midair mating ritual, one eagle flips upside down and locks talons with another. Both tumble, then break apart to continue flight. Not uncommon for juveniles to perform this mating ritual even though they are not yet breeding age. Juveniles attain the white head and tail at 4–5 years of age.

in flight

Wood Stork

Mycteria americana

SUMMER MIGRATION

Size: 42–44" (107–112 cm); up to 5' wingspan

Male: An all-white body with a bald, nearly black head. Tail, wing tips and entire trailing edge of wings are black, as seen in flight. Black legs and pink feet. Thick, slightly down-curved dark bill.

Female: same as male

Juvenile: similar to adult but with a gray-to-brown head and neck and a dull-yellow bill

Nest: platform; female and male build; 1 brood per year

Eggs: 2–4; white without markings

Incubation: 28–32 days; male and female incubate

Fledging: 55–60 days; female and male feed the young

Migration: complete, to Florida

Food: fish, amphibians, aquatic insects, snails

Compare: Great Egret (p. 355) is similarly sized but lacks Wood Stork's dark head and down-curved bill. Snowy Egret (p. 349) is about half the size of Wood Stork. Look for Wood Stork's black tail, wing tips and edge of wings in flight.

Stan's Notes: On the federal endangered species list. Like many other wading birds, it has less than 20 percent of the population it did a century ago. Feeds by swinging its open bill through water until it contacts prey, then snaps bill shut. Shuffles its feet to stir up fish before capturing. Nests in large colonies, often high up in trees. Usually doesn't breed until it reaches 4–5 years of age. Abandons eggs or young when food supply is short.

Blue-gray Gnatcatcher

Polioptila caerulea

YEAR-ROUND
SUMMER
WINTER

Size: 4" (10 cm)

Male: A light-blue-to-gray head, back, breast and wings, with a white belly. Black forehead and eyebrows. Prominent white eye-ring. Long black tail with a white undertail, often held cocked above the rest of body.

Female: same as male but grayer and lacking black on the head

Juvenile: similar to female

Nest: cup; female and male construct; 1 brood per year

Eggs: 4–5; pale blue with dark markings

Incubation: 10–13 days; female and male incubate

Fledging: 10–12 days; female and male feed the young

Migration: complete to non-migrator in Louisiana and Mississippi

Food: insects

Compare: The only small blue bird with a black tail. Very active near the nest, look for it flitting around upper branches in search of insects.

Stan's Notes: Found in a wide variety of forest types throughout Louisiana and Mississippi. Listen for its wheezy call notes to help locate it. Flicks its tail up and down and from side to side while calling. In many years, it nests so early that by mid-June it is no longer defending territory. Like many open-woodland nesters, it is a common cowbird host. Although the population is abundant and widespread, it has been decreasing in the recent past. Many northern Blue-gray Gnatcatchers winter in southern Louisiana and Mississippi, swelling local populations.

female
p. 137

male

Indigo Bunting

Passerina cyanea

**YEAR-ROUND
SUMMER**

Size: 5½" (14 cm)

Male: Vibrant-blue finch-like bird. Dark markings scattered on wings and tail.

Female: light-brown with faint markings

Juvenile: similar to female

Nest: cup; female builds; 2 broods per year

Eggs: 3–4; pale blue without markings

Incubation: 12–13 days; female incubates

Fledging: 10–11 days; female feeds the young

Migration: complete, to Mexico, Central America and South America; a few non-migrators in southern Louisiana

Food: insects, seeds, fruit; will visit seed feeders

Compare: The male Eastern Bluebird (p. 109) is larger and has a rust-red chest. Male Blue Grosbeak (p. 107) is larger, has chestnut-colored wing bars and a large bill. Look for the bright-blue plumage to identify the male Indigo Bunting.

Stan's Notes: Seen along woodland edges and in parks and yards, feeding on insects. Comes to seed feeders early in spring, before insects are plentiful. Usually only the males are noticed. The male often sings from treetops to attract a mate. The female is quiet. Actually a gray bird, without blue pigment in its feathers: like Blue Jays and other blue birds, sunlight is refracted within the structure of the feathers, making them appear blue. Plumage is iridescent in direct sun, duller in shade. Molts in spring to acquire body feathers with gray tips, which quickly wear off, revealing the bright-blue plumage. Molts in fall and appears like the female during winter. Migrates at night in flocks of 5–10 individuals.

Tree Swallow
Tachycineta bicolor

SUMMER
MIGRATION
WINTER

Size: 5–6" (13–15 cm)

Male: Blue-green in spring, greener in fall. Changes color in direct sunlight. White from chin to belly. Long, pointed wing tips. Notched tail.

Female: similar to male but duller

Juvenile: gray brown with a white belly and a grayish breast band

Nest: cavity; female and male line a vacant woodpecker cavity or nest box; 2 broods per year

Eggs: 4–6; white without markings

Incubation: 13–16 days; female incubates

Fledging: 20–24 days; female and male feed the young

Migration: complete to Louisiana, Mississippi, other southern coastal states, Mexico and Central America

Food: insects

Compare: The Purple Martin (p. 111) is much larger and darker. The Barn Swallow (p. 105) has a rusty belly and a long, deeply forked tail. Look for the white chin, chest and belly and the notched tail to help identify the Tree Swallow.

Stan's Notes: Most common at coastal beaches, freshwater ponds, lakes and agricultural fields. Can be attracted to your yard with a nest box. Competes with Eastern Bluebirds for cavities and nest boxes. Builds a grass nest within and will travel long distances, looking for dropped feathers for the lining. Watch for it playing and chasing after feathers. Often seen flying back and forth across fields, feeding on insects. Eats many nuisance bugs. Gathers in large flocks during migration and winter.

Barn Swallow
Hirundo rustica

SUMMER

Size: 7" (18 cm)

Male: Sleek swallow. Blue-black back, cinnamon belly and reddish-brown chin. White spots on a long, deeply forked tail.

Female: same as male but with a whitish belly

Juvenile: similar to adults, with a tan belly and chin, and shorter tail

Nest: cup; female and male build; 2 broods per year

Eggs: 4–5; white with brown markings

Incubation: 13–17 days; female incubates

Fledging: 18–23 days; female and male feed the young

Migration: complete, to South America

Food: insects (prefers beetles, wasps, flies)

Compare: The Tree Swallow (p. 103) is white from chin to belly and has a notched tail. The Purple Martin (p. 111) is larger and has a dark-purple belly. The Chimney Swift (p. 125) has a narrow, pointed tail. Look for the deeply forked tail to identify the Barn Swallow.

Stan's Notes: Seen in wetlands, farms, suburban yards and parks. Of the six swallow species in Louisiana and Mississippi, this is the only one with a deeply forked tail. Unlike other swallows, it rarely glides in flight. Usually flies low over land or water. Drinks as it flies, skimming water, or will sip water droplets on wet leaves. Bathes while flying through rain or sprinklers. Gives a twittering warble, followed by a mechanical sound. Builds a mud nest with up to 1,000 beak-loads of mud. Nests on barns and houses, under bridges and in other sheltered places. Often nests in colonies of 4–6 birds; sometimes nests alone.

male

female
p. 157

Blue Grosbeak
Passerina caerulea

SUMMER
MIGRATION

Size: 7" (18 cm)

Male: Overall blue bird with 2 chestnut wing bars. Large gray-to-silver bill. Black around base of bill.

Female: overall brown with darker wings and tail, 2 tan wing bars, large gray-to-silver bill

Juvenile: similar to female

Nest: cup; female builds; 1–2 broods per year

Eggs: 3–6; pale blue without markings

Incubation: 11–12 days; female incubates

Fledging: 9–10 days; female and male feed the young

Migration: complete, to the Bahamas, Cuba, Mexico and Central America

Food: insects, seeds; will come to seed feeders

Compare: The more common male Indigo Bunting (p. 101) is very similar, but it is smaller and lacks wing bars. The male Eastern Bluebird (p. 109) is the same size, but it lacks the wing bars and oversized bill.

Stan's Notes: A common summer resident and migrant. Returns to Louisiana and Mississippi by mid-April and leaves in October. It has expanded northward, and its overall populations have increased over the past 30–40 years. A bird of semi-open habitats such as overgrown fields, riversides, woodland edges and fence-rows. Visits seed feeders, where it is often confused with male Indigo Buntings. Frequently seen twitching and spreading its tail. The first-year males show only some blue, obtaining the full complement of blue feathers in the second winter.

Eastern Bluebird
Sialia sialis

YEAR-ROUND

Size:	7" (18 cm)
Male:	Sky-blue head, back and tail. Rust-red breast and white belly.
Female:	grayer than male, with a faint rusty breast and faint blue wings and tail
Juvenile:	similar to female but with spots on the breast and blue wing markings
Nest:	cavity, vacant woodpecker cavity or nest box; female adds a soft lining; 2 broods per year
Eggs:	4–5; pale blue without markings
Incubation:	12–14 days; female incubates
Fledging:	15–18 days; male and female feed the young
Migration:	non-migrator in Louisiana and Mississippi
Food:	insects, fruit; comes to shallow dishes with live or dead mealworms, and to suet feeders
Compare:	The male Indigo Bunting (p. 101) is nearly all blue. The Blue Jay (p. 113) is much larger and has a crest. Look for the rusty breast to help identify the Eastern Bluebird.

Stan's Notes: Once nearly eliminated from Louisiana and Mississippi due to a lack of nest cavities. Thanks to people who installed thousands of nest boxes, bluebirds now thrive. Prefers open habitats, such as farm fields, pastures and roadsides, but also likes forest edges, parks and yards. Often perches on trees or fence posts and drops to the ground to grab bugs, especially grasshoppers. Makes short flights from tree to tree. Song is a distinctive "churlee chur chur-lee." A year-round resident that is joined by many northern migrants, swelling populations during winter. The rust-red breast is like that of the American Robin, its cousin.

Purple Martin

Progne subis

Size: 8½" (22 cm)

Male: Iridescent with a purple-to-black head, back and belly. Black wings and a notched black tail.

Female: grayish-purple head and back, darker wings and tail, whitish belly

Juvenile: same as female

Nest: cavity; female and male line the cavity of the house; 1 brood per year

Eggs: 4–5; white without markings

Incubation: 15–18 days; female incubates

Fledging: 26–30 days; male and female feed the young

Migration: complete, to South America

Food: insects

Compare: Usually seen only in groups. The male Purple Martin is the only swallow with a very dark-purplish belly.

SUMMER

Stan's Notes: The largest swallow species in North America. Once nested in tree cavities; now nests almost exclusively in man-made, apartment-style houses. The most successful colonies often nest in multiunit nest boxes within 100 feet (30 m) of a human dwelling near a lake. Main diet consists of dragonflies, not mosquitoes, as once thought. Gives a continuous stream of chirps, creaks and rattles, along with a shout-like "churrr" and chortle. Often drinks in flight, skimming water, and bathes in flight, flying through rain. Returns to the same nest site each year; the males arrive before the females and yearlings. The young leave to form new colonies. Large colonies gather in fall before migrating to South America.

Blue Jay
Cyanocitta cristata

YEAR-ROUND

Size: 12" (30 cm)

Male: Bright light-blue-and-white bird with a black necklace and gray belly. Large crest moves up and down at will. White face, wing bars and tip of tail. Black tail bands.

Female: same as male

Juvenile: same as adult but duller

Nest: cup; female and male construct; 1–2 broods per year

Eggs: 4–5; green to blue with brown markings

Incubation: 16–18 days; female incubates

Fledging: 17–21 days; female and male feed the young

Migration: non-migrator to partial migrator; will move around to find an abundant food source

Food: insects, fruit, carrion, seeds, nuts; visits seed feeders, ground feeders with corn or peanuts

Compare: The Belted Kingfisher (p. 115) has a larger, more ragged crest. The Eastern Bluebird (p. 109) is much smaller and has a rust-red breast. Look for the large crest to help identify the Blue Jay.

Stan's Notes: Highly intelligent, solving problems, gathering food and communicating more than other birds. Loud and noisy; mimics other birds. Known as the alarm of the forest, screaming at intruders. Imitates hawk calls around feeders to scare off other birds. One of the few birds to cache food; can remember where it hid thousands of nuts. Carries food in a pouch under its tongue (sublingually). Eats eggs and young birds from other nests. Feathers lack blue pigment; refracted sunlight causes the blue appearance.

YEAR-ROUND

Belted Kingfisher
Megaceryle alcyon

Size: 12–14" (30–36 cm)

Male: Blue with white belly, blue-gray chest band, and black wing tips. Ragged crest moves up and down at will. Large head. Long, thick, black bill. White spot by eyes. Red-brown eyes.

Female: same as male but with rusty flanks and a rusty chest band below the blue-gray band

Juvenile: similar to female

Nest: cavity; female and male excavate in a bank of a river, lake or cliff; 1 brood per year

Eggs: 6–7; white without markings

Incubation: 23–24 days; female and male incubate

Fledging: 23–24 days; female and male feed the young

Migration: non-migrator in Louisiana and Mississippi

Food: small fish

Compare: The Blue Jay (p. 113) is lighter blue and has a plain gray chest and belly. The Belted Kingfisher is rarely found away from water.

Stan's Notes: Usually found at the bank of a river, lake or large stream. Perches on a branch near water, dives in headfirst to catch a small fish, then returns to the branch to feed. Parents drop dead fish into the water to teach their young to dive. Can't pass bones through its digestive tract; regurgitates bone pellets after meals. Gives a loud call that sounds like a machine gun. Mates know each other by their calls. Digs a tunnel up to 4 feet (1 m) long to a nest chamber. Small white patches on dark wing tips flash during flight. Resident birds are joined by migrants from northern states, swelling population in winter.

SUMMER

Purple Gallinule
Porphyrio martinicus

Size: 13" (33 cm)

Male: A vibrant blue head, breast and belly with iridescent green back and wings. Yellow-tipped red bill. White undertail. Yellow legs.

Female: same as male

Juvenile: brown version of adult; bronze legs

Nest: ground; female and male build; 1–2 broods per year

Eggs: 6–8; brown with dark markings

Incubation: 22–25 days; female and male incubate

Fledging: 55–60 days; female and male feed the young

Migration: complete, to Florida, Mexico and Central America

Food: insects, snails, seeds, berries, frogs

Compare: The American Coot (p. 37) is similar in size with a yellow-tipped red bill. Common Gallinule (p. 35) is similar in size, but it has a white side stripe that the Purple Gallinule lacks. Look for white undertail to help identify the Purple Gallinule.

Stan's Notes: This is one of the most dramatic-looking birds in Louisiana and Mississippi. Uses its extremely long toes to walk on floating vegetation in freshwater and saltwater marshes, where it hunts for grasshoppers and other insects, grains and frogs. Family groups stay together; first brood sometimes helps raise the second. Individuals have been known to wander well north of Louisiana and Mississippi.

non-breeding

breeding

molting juvenile

white juvenile

Little Blue Heron
Egretta caerulea

YEAR-ROUND
SUMMER
MIGRATION

Size: 22–26" (56–66 cm)

Male: Dark slate-blue to purple nearly all year. Dull-green legs and feet. Black-tipped blue-gray bill. Breeding adult has a reddish-purple head and neck with several long plumes on the crown.

Female: same as male

Juvenile: pure white overall, yellowish legs and feet, black-tipped gray bill

Nest: platform; female and male build; 1 brood

Eggs: 2–6; light blue without markings

Incubation: 20–23 days; female and male incubate

Fledging: 42–49 days; female and male feed the young

Migration: partial to non-migrator in Louisiana and Mississippi

Food: fish, aquatic insects

Compare: Tricolored Heron (p. 121) has a white belly. Snowy Egret (p. 349) can be confused with a juvenile Little Blue, but Snowy has bright-yellow feet, black legs and a solid black bill. Cattle Egret (p. 347) has an orange-buff crest, breast and back, and a red-orange bill.

Stan's Notes: Often flies north after breeding season, returning to southern Louisiana and Mississippi for the winter. Unusual in that the young look completely different from adults. All-white young turn blotchy white the first year. By the second year they look like the adult birds. A very slow stalker of prey in freshwater lakes and rivers, saltwater marshes and wetlands. Nests in large colonies near saltwater sites.

breeding

non-breeding

Tricolored Heron

Egretta tricolor

YEAR-ROUND
SUMMER
MIGRATION

Size: 24–28" (60–71 cm)

Male: Dark-blue head, wings and back of neck. White belly and white stripe on underside of neck. Small brown patches at base of neck, with lighter brown on the lower back. Legs are yellow to pale green. Long, slender bill with a dark tip. The non-breeding adult Tricolored Heron lacks a blue face and base of bill.

Female: same as male

Juvenile: similar to adult, but chestnut-brown in place of dark-blue areas

Nest: platform; female and male build; 1 brood per year

Eggs: 3–6; light blue without markings

Incubation: 21–25 days; female and male incubate

Fledging: 32–35 days; female and male feed the young

Migration: partial to non-migrator in southern Louisiana and Mississippi

Food: fish, aquatic insects

Compare: Great Blue Heron (p. 305) is much larger and lacks white undersides. The Little Blue Heron (p. 119) is smaller and lacks white on neck.

Stan's Notes: A medium-size heron characterized by its white undersides. Like other herons, numbers have declined due to habitat loss. A year-round resident, although less numerous in the winter. Seen mainly in saltwater marshes and estuaries, but also in freshwater marshes inland. In spring and summer, known to wander as far as the northern tier states. Colony nester with other herons, one adult always on duty at the nest.

white
morph

dark morph

hunting

Reddish Egret
Egretta rufescens

YEAR-ROUND

Size: 30" (76 cm)

Male: A slate blue-bodied egret with a shaggy reddish head and neck. Long dark legs and feet. Long, pointed black-tipped bill. White morph is all white with shaggy plumes at base of neck and a dark-tipped pink bill.

Female: same as male

Juvenile: pale version of adult

Nest: platform; female and male build; 1 brood per year

Eggs: 3–4; light blue without markings

Incubation: 25–26 days; female and male incubate

Fledging: 42–46 days; female and male feed the young

Migration: non-migrator

Food: fish, aquatic insects

Compare: Tricolored Heron (p. 121) is smaller, has a white belly and lacks the reddish head and neck. The Little Blue Heron (p. 119) has a blue-gray bill with a dark tip. Snowy Egret (p. 349) has yellow feet and a yellow mark at the base of its bill.

Stan's Notes: Inhabits saltwater on the Louisiana coast. "Dances" on shallow water, hunting. Two color morphs, with the dark morph more common that the white. Identified by its unusual hunting behavior. Runs with wings wide open to shade water, darting its head to grab fish. May stir the bottom with its feet to expose fish. Nearly eliminated by plume hunters by the late 1800s. Populations have rebounded, but the species remains uncommon. Often alone and silent. May croak or bark upon takeoff when disturbed.

Chimney Swift
Chaetura pelagica

SUMMER

Size: 5" (13 cm)

Male: Nondescript, cigar-shaped bird, usually seen in flight. Long, thin, brown body. Pointed tail and head. Long, backswept wings, longer than the body.

Female: same as male

Juvenile: same as adult

Nest: half cup; female and male build; 1 brood per year

Eggs: 4–5; white without markings

Incubation: 19–21 days; female and male incubate

Fledging: 28–30 days; female and male feed the young

Migration: complete, to South America

Food: insects caught in midair

Compare: The Purple Martin (p. 111) is much larger and darker. The Barn Swallow (p. 105) has a deeply forked tail. Tree Swallows (p. 103) have a white belly and blue-green back.

Stan's Notes: One of the fastest fliers in the bird world. Spends all day flying, rarely perching. Flies in groups, feeding on insects flying 100 feet (30 m) or higher up in the air. Often called a Flying Cigar due to its body shape, which is pointed at both ends. Drinks and bathes during flight, skimming water. Gives a unique in-flight twittering call, often heard before the bird is seen. Hundreds roost in large chimneys, giving it the common name. Builds its nest with tiny twigs, cementing it with saliva and attaching it to the inside of a chimney or a hollow tree. Hundreds roost in large chimneys, hence the common name. Nested in hollow trees before chimneys; now most nest in man-made structures. Usually only one nest per chimney.

Chipping Sparrow
Spizella passerina

YEAR-ROUND

Size: 5" (13 cm)

Male: Small gray-brown sparrow with clear-gray chest. Rusty crown. White eyebrows and thin black eye line. Thin gray-black bill. Two faint wing bars.

Female: same as male

Juvenile: similar to adults, with streaking on the chest; lacks a rusty crown

Nest: cup; female builds; 2 broods per year

Eggs: 3–5; blue-green with brown markings

Incubation: 11–14 days; female incubates

Fledging: 10–12 days; female and male feed the young

Migration: non-migrator in Louisiana and Mississippi

Food: insects, seeds; will come to ground feeders

Compare: The Song Sparrow (p. 141) and female House Finch (p. 131) have heavily streaked chests. Look for the rusty crown and black eye line to help identify the Chipping Sparrow.

Stan's Notes: A year-round resident throughout Louisiana and Mississippi, common in gardens and yards. Many northern birds join these residents in the winter. Often seen feeding on dropped seeds beneath feeders. In northern states, gathers in large family groups to feed in preparation for migration. Migrates at night in flocks of 20–30 birds. The common name comes from the male's fast "chip" call. Often is just called Chippy. Builds nest low in dense shrubs and almost always lines it with animal hair. Comfortable with people, allowing you to approach closely before it flies away.

Pine Siskin

Spinus pinus

WINTER

Size: 5" (13 cm)

Male: Small brown finch with heavy streaking on the back, breast and belly. Yellow wing bars. Yellow at the base of the tail. Thin bill.

Female: similar to male, with less yellow

Juvenile: similar to adult, with a light-yellow tinge over the breast and chin

Nest: modified cup; female builds; 2 broods

Eggs: 3–4; greenish blue with brown markings

Incubation: 12–13 days; female incubates

Fledging: 14–15 days; female and male feed the young

Migration: irruptive; moves around the United States in search of food during winter

Food: seeds, insects; will come to seed feeders

Compare: The female Purple Finch (p. 145) has white eyebrows. The female House Finch (p. 131) lacks yellow. The female American Goldfinch (p. 359) has white wing bars. Look for the yellow wing bars to identify the Pine Siskin.

Stan's Notes: This bird is usually considered a winter finch. Can be found throughout Louisiana and Mississippi during heavy invasion years, but is absent in many winters. Seen in flocks of up to 20 birds, often with other finch species. Gathers in flocks and moves around, visiting feeders. Gives a series of high-pitched, wheezy calls. Builds nest toward the end of coniferous branches, where needles are dense, helping to conceal. Nests are often only a few feet apart. Male feeds the female during incubation. Juveniles lose the yellow tint by late summer of their first year. Adult males have more yellow on their wings than the adult females.

male
p. 327

female

House Finch
Haemorhous mexicanus

YEAR-ROUND

Size: 5" (13 cm)

Female: Plain brown with heavy streaking on a white chest.

Male: red-to-orange face, throat, chest and rump; streaked belly and wings; brown cap; brown marking behind the eyes

Juvenile: similar to female

Nest: cup, occasionally in a cavity; female builds; 2 broods per year

Eggs: 4–5; pale blue, lightly marked

Incubation: 12–14 days; female incubates

Fledging: 15–19 days; female and male feed the young

Migration: non-migrator; will move around to find food

Food: seeds, fruit, leaf buds; visits seed feeders and feeders that offer grape jelly

Compare: The female Purple Finch (p. 145) has bold white eyebrows. The Pine Siskin (p. 129) has yellow wing bars and a smaller bill. The female American Goldfinch (p. 359) has a clear chest. Look for the heavily streaked chest to help identify the female House Finch.

Stan's Notes: Can be a common bird at your feeders. A very social bird, visiting feeders in small flocks. Likes to nest in hanging flower baskets. Incubating female is fed by the male. Male sings a loud, cheerful warbling song. It was originally introduced to Long Island, New York, from the western U.S. in the 1940s. Now found throughout the country. Suffers from a disease that causes the eyes to crust, resulting in blindness and death.

House Wren
Troglodytes aedon

MIGRATION
WINTER

Size: 5" (13 cm)

Male: All-brown bird with lighter-brown markings on the wings and tail. Slightly curved brown bill. Often holds tail upward.

Female: same as male

Juvenile: same as adult

Nest: cavity; female and male line just about any nest cavity; 2 broods per year

Eggs: 4–6; tan with brown markings

Incubation: 10–13 days; female and male incubate

Fledging: 12–15 days; female and male feed the young

Migration: complete, to Louisiana, Mississippi, other southern states and Mexico

Food: insects, spiders, snails

Compare: Carolina Wren (p. 135) has prominent eyebrows. Look for House Wren's long curved bill and long upturned tail to differentiate it from sparrows.

Stan's Notes: A prolific songster. During the mating season, sings from dawn to dusk. Seen in brushy yards, parks and woodlands and along forest edges. Easily attracted to a nest box. In spring, the male chooses several prospective nesting cavities and places a few small twigs in each. The female inspects all of them and finishes constructing the nest in the cavity of her choice. She fills the cavity with short twigs and then lines a small depression at the back with pine needles and grass. She often has trouble fitting longer twigs through the entrance hole and tries many different directions and approaches until she is successful. A winter resident and migrator in Louisiana and Mississippi.

Carolina Wren

Thryothorus ludovicianus

Size: 5½" (14 cm)

Male: Rusty-brown head and back with an orange-yellow chest and belly. White throat and a prominent white eye stripe. Short, stubby tail, often cocked up.

Female: same as male

Juvenile: same as adults

Nest: cavity; female and male build; 2 broods per year, sometimes 3

Eggs: 4–6; white, sometimes pink or creamy, with brown markings

Incubation: 12–14 days; female incubates

Fledging: 12–14 days; female and male feed the young

Migration: non-migrator; moves around in winter

Food: insects, fruit, few seeds; visits suet feeders

Compare: House Wren (p. 133) is darker brown and lacks a white eye stripe.

Stan's Notes: Year-round resident in Louisiana and Mississippi. Mates are long-term, staying together throughout the year in permanent territories. Sings year-round. The male is known to sing up to 40 song types, singing one song repeatedly before switching to another. The female also sings, resulting in duets. The male often takes over feeding the first brood while the female renests. Nests in birdhouses and in unusual places like mailboxes, bumpers or broken taillights of vehicles, or nearly any other cavity. Found in brushy yards or woodlands. Can be attracted to feeders with mealworms.

female

male
p. 101

Indigo Bunting
Passerina cyanea

YEAR-ROUND
SUMMER

Size: 5½" (14 cm)

Female: Light-brown, finch-like bird. Faint streaking on a light-tan chest. Wings have a very faint blue cast and indistinct wing bars.

Male: vibrant blue with scattered dark markings on wings and tail

Juvenile: similar to female

Nest: cup; female builds; 2 broods per year

Eggs: 3–4; pale blue without markings

Incubation: 12–13 days; female incubates

Fledging: 10–11 days; female feeds the young

Migration: complete, to Mexico, Central America and South America; a few non-migrators in southern Louisiana

Food: insects, seeds, fruit; will visit seed feeders

Compare: Female Blue Grosbeak (p. 157) is larger and has 2 tan wing bars. The female Purple Finch (p. 145) has white eyebrows and heavy streaking on the chest. The female House Finch (p. 131) has a heavily streaked chest. The female American Goldfinch (p. 359) has white wing bars. Look for the faint blue cast on the wings to help identify the female Indigo Bunting.

Stan's Notes: Seen along woodland edges and in parks and yards, feeding on insects. Comes to seed feeders early in spring, before insects are plentiful. Secretive, plain and quiet; usually only the males are noticed. The male often sings from treetops to attract a mate. Migrates at night in flocks of 5–10 birds. Males return before the females and juveniles, often to the nest site of the last year.

137

male
p. 255

female

Dark-eyed Junco

Junco hyemalis

WINTER

Size: 5½" (14 cm)

Female: Plump, dark-eyed bird with a tan-to-brown chest, head and back. White belly. Ivory-to-pink bill. White outer tail feathers appear like a white V in flight.

Male: round with gray plumage

Juvenile: similar to female, with streaking on the breast and head

Nest: cup; female and male build; 2 broods per year

Eggs: 3–5; white with reddish-brown markings

Incubation: 12–13 days; female incubates

Fledging: 10–13 days; male and female feed the young

Migration: complete to Louisiana and Mississippi and other southern states

Food: seeds, insects; visits ground and seed feeders

Compare: Rarely confused with any other bird. Look for the ivory-to-pink bill and small flocks feeding under feeders to help identify the female Dark-eyed Junco.

Stan's Notes: A common winter bird in Louisiana and Mississippi. Migrates from Canada to Louisiana and Mississippi and beyond. Females tend to migrate farther south than males. Adheres to a rigid social hierarchy, with dominant birds chasing the less dominant ones. Look for the white outer tail feathers flashing in flight. Often seen in small flocks on the ground, where it uses its feet to simultaneously "double-scratch" to expose seeds and insects. Eats many weed seeds. Nests in a wide variety of wooded habitats. Several subspecies of Dark-eyed Junco were previously considered to be separate species.

Song Sparrow
Melospiza melodia

YEAR-ROUND
WINTER

Size: 5–6" (13–15 cm)

Male: Common brown sparrow with heavy dark streaks on the chest coalescing into a central dark spot.

Female: same as male

Juvenile: similar to adults, with a finely streaked chest; lacks a central dark spot

Nest: cup; female builds; 2 broods per year

Eggs: 3–4; blue to green, with red-brown markings

Incubation: 12–14 days; female incubates

Fledging: 9–12 days; female and male feed the young

Migration: complete, to Louisiana and Mississippi and other southern states

Food: insects, seeds; only rarely comes to ground feeders with seeds

Compare: Similar to other brown sparrows. Look for the heavily streaked chest with a central dark spot to help identify the Song Sparrow.

Stan's Notes: There are many subspecies of this bird, but the dark spot in the center of the chest appears in every variety. A constant songster, repeating its loud, clear song every few minutes. The song varies from region to region but has the same basic structure. Sings from thick shrubs to defend a small territory, beginning with three notes and finishing up with a trill. A ground feeder, it will "double-scratch" with both feet at the same time to expose seeds. When the female builds a new nest for a second brood, the male often takes over feeding the first brood. Unlike many other sparrow species, Song Sparrows rarely flock together. A common host of the Brown-headed Cowbird.

male

female

House Sparrow

Passer domesticus

YEAR-ROUND

Size: 6" (15 cm)

Male: Brown back with a gray belly and cap. Large black patch extending from the throat to the chest (bib). One white wing bar.

Female: slightly smaller than the male; light brown with light eyebrows; lacks a bib and white wing bar

Juvenile: similar to female

Nest: cavity; female and male build a domed cup nest within; 2–3 broods per year

Eggs: 4–6; white with brown markings

Incubation: 10–12 days; female incubates

Fledging: 14–17 days; female and male feed the young

Migration: non-migrator; moves around to find food

Food: seeds, insects, fruit; comes to seed feeders

Compare: Chipping Sparrow (p. 127) has a rusty crown. Look for the black bib to identify the male House Sparrow and the clear breast to help identify the female.

Stan's Notes: One of the first birdsongs heard in cities in spring. A familiar city bird, nearly always in small flocks. Also found on farms. Introduced in 1850 from Europe to Central Park in New York. Now seen throughout North America. Related to old-world sparrows; not a relative of any sparrows in the U.S. An aggressive bird that will kill young birds in order to take over the nest cavity. Uses dried grass and small scraps of plastic, paper and other materials to build an oversize, domed nest in the cavity.

male
p. 329

female

Purple Finch
Haemorhous purpureus

WINTER

Size: 6" (15 cm)

Female: Plain brown with heavy streaking on the chest. Bold white eyebrows and a large bill.

Male: raspberry-red head, cap, breast, back and rump; brownish wings and tail

Juvenile: same as female

Nest: cup; female and male build; 1 brood per year

Eggs: 4–5; greenish blue with brown markings

Incubation: 12–13 days; female incubates

Fledging: 13–14 days; female and male feed the young

Migration: irruptive; moves around in search of food

Food: seeds, insects, fruit; comes to seed feeders

Compare: The female House Finch (p. 131) lacks eyebrows. Pine Siskin (p. 129) has yellow wing bars. The female American Goldfinch (p. 359) has a clear chest. Look for the bold white eyebrows to identify the female Purple Finch.

Stan's Notes: Found in Louisiana and Mississippi in winter when flocks of Purple Finches leave their homes farther north and move around searching for food. Seen in some winters, absent in others. Travels in flocks of up to 50 birds. Visits seed feeders along with House Finches, which makes it hard to tell them apart. Feeds mainly on seeds; ash tree seeds are an important source of food. Found in coniferous forests, mixed woods, woodland edges and suburban backyards. Flies in the typical undulating, up-and-down pattern of finches. Sings a rich, loud song. Gives a distinctive "tic" note only in flight. The male is not purple. The Latin species name *purpureus* means "purple" (and other reddish colors).

Louisiana Waterthrush
Parkesia motacilla

SUMMER
MIGRATION

Size: 6" (15 cm)

Male: Thin, dark-bodied bird with bold white eyebrows. A white chin, breast and belly with large vertical dark streaks. Short squared-off tail, often cocked upward. Thin pink legs and feet. A thin, pointed dark bill.

Female: same as male

Juvenile: similar to adult

Nest: cup; female builds; 1 brood per year

Eggs: 4–6; creamy white with dark markings

Incubation: 13–15 days; female incubates

Fledging: 10–12 days; female and male feed the young

Migration: complete, to Mexico, Central and South America, the Caribbean

Food: insects, mollusks, crustaceans, tiny fish

Compare: The female Purple Finch (p. 145) has similar white eyebrows and a streaked breast, but is seen during winter at bird feeders, while the Louisiana Waterthrush is seen during spring and summer along streams.

Stan's Notes: An unusual warbler that needs fast-moving streams and river swamps. Runs along the shore or on partially submerged rocks, looking for bugs. One of the first warblers to arrive in spring and leave in fall. Builds a well-hidden cup nest along a stream bank, often beneath an uprooted tree or overhanging rock by the water. May build a walkway of leaves to and from the water's edge. Unlike other warblers, walks on the ground, bobbing its head and tail. Has a beautiful clear whistle that can be heard over running water.

winter
p. 259

breeding

Least Sandpiper
Calidris minutilla

Size: 6" (15 cm)

Male: Breeding plumage has a golden-brown head and back and a white belly. Dull-yellow legs. White eyebrows and a short, down-curved black bill.

Female: same as male

Juvenile: similar to winter adult but buff-brown and lacks the breast band

Nest: ground; male and female construct; 1 brood per year

Eggs: 3–4; olive with dark markings

Incubation: 19–23 days; male and female incubate

Fledging: 25–28 days; male and female feed the young

Migration: complete, to Louisiana, Mississippi, other southern coastal states, Mexico, and Central and South America

Food: aquatic and terrestrial insects, seeds

Compare: The smallest of sandpipers. Often confused with breeding Western Sandpiper (p. 151). Look for Least Sandpiper's yellow legs to differentiate it from other tiny sandpipers. The short, thin down-curved bill also helps to identify.

Stan's Notes: A winter resident across Louisiana and Mississippi. This is a tiny, tame sandpiper that can be approached without scaring it. It is the smallest of peeps (sandpipers), nesting on the tundra in northern regions of Canada and Alaska. Prefers the grassy flats of saltwater and freshwater ponds. Its yellow legs can be hard to see in water, poor light or when covered with mud. Most other small shorebirds have black legs and feet.

winter
p. 261

breeding

MIGRATION
WINTER

Western Sandpiper
Calidris mauri

Size: 6½" (16 cm)

Male: Breeding has a bright rust-brown crown, ear patch and back and a white chin and chest. Black legs. Narrow bill that droops near tip.

Female: same as male

Juvenile: similar to breeding adult; bright buff-brown on the back only

Nest: ground; male and female construct; 1 brood per year

Eggs: 2–4; light brown with dark markings

Incubation: 20–22 days; male and female incubate

Fledging: 19–21 days; male and female feed the young

Migration: complete, to Louisiana, Mississippi, other southern coastal states, Mexico, and Central and South America

Food: aquatic and terrestrial insects

Compare: Breeding Least Sandpiper (p. 149) lacks the bright rust-brown cap, ear patch and back. Least Sandpiper has yellow legs. Look for Western's longer bill that droops slightly.

Stan's Notes: A winter resident and migrator across Louisiana and Mississippi. Winters in coastal states from Delaware to California. Nests on the ground in large "loose" colonies on the tundra of northern coastal Alaska. Adults leave their breeding grounds several weeks before the young. Some obtain their breeding plumage before leaving in spring. Feeds on insects at the water's edge, sometimes immersing its head. Young leave the nest (precocial) within a few hours after hatching. Female leaves and the male tends the hatchlings.

white-striped

tan-striped

WINTER

White-throated Sparrow
Zonotrichia albicollis

Size: 6–7" (15–18 cm)

Male: Brown with a gray or tan chest and belly. Bold striping on the head. White or tan throat patch and eyebrows. Small yellow spot in the space between the eye and bill, called the lore.

Female: same as male

Juvenile: similar to adults, with a heavily streaked chest and a gray throat and eyebrows

Nest: cup; female builds; 1 brood per year

Eggs: 4–6; green to blue, or creamy white with red-brown markings

Incubation: 11–14 days; female incubates

Fledging: 10–12 days; female and male feed the young

Migration: complete, to Louisiana and Mississippi, other southern states and Mexico

Food: insects, seeds, fruit; visits ground feeders

Compare: Song Sparrow (p. 141) has a central spot on chest and lacks a striped pattern on head. Whited-crowned Sparrow (p. 155) lacks a throat patch and yellow lore.

Stan's Notes: Two color variations (polymorphic): white striped and tan striped. Studies indicate that the white-striped adults tend to mate with the tan-striped birds; it's not clear why. Known for its wonderful song; it sings all year and can even be heard at night. White- and tan-striped males and white-striped females sing, but tan-striped females do not. Feeds on the ground under feeders. Builds nest on the ground under small trees in bogs and coniferous forests. Often associated with other sparrows in winter.

juvenile

White-crowned Sparrow
Zonotrichia leucophrys

WINTER

Size: 6½–7½" (16.5–19 cm)

Male: Brown with a gray chest and black-and-white striped crown. Small, thin pink bill.

Female: same as male

Juvenile: similar to adults, with black and brown stripes on the head

Nest: cup; female builds; 2 broods per year

Eggs: 3–5; greenish to bluish to whitish with red-brown markings

Incubation: 11–14 days; female incubates

Fledging: 8–12 days; male and female feed the young

Migration: complete, to Louisiana, Mississippi, other southern states and Mexico

Food: insects, seeds, berries; visits ground feeders

Compare: The White-throated Sparrow (p. 153) has a throat patch and a small yellow spot by each eye. The Song Sparrow (p. 141) has a streaked chest. Look for the striped crown to help identify the White-crowned Sparrow.

Stan's Notes: Often seen in groups of up to 20 birds during migration and winter, when it can be seen visiting ground feeders and feeding beneath seed feeders. This ground feeder will "double-scratch" backward with both feet simultaneously to find seeds. Prefers scrubby areas, woodland edges, and open or grassy habitats. Males arrive at the breeding grounds before the females and sing from perches to establish territory. Males take most of the responsibility for raising the young while females start their second broods. Only 9–12 days separate the broods. Nests in Canada and Alaska.

male
p. 107

female

Blue Grosbeak
Passerina caerulea

SUMMER
MIGRATION

Size: 7" (18 cm)

Female: Overall brown with darker wings and tail. Two tan wing bars. Large gray-to-silver bill.

Male: blue bird with 2 chestnut wing bars; large gray-to-silver bill; black around base of bill

Juvenile: similar to female

Nest: cup; female builds; 1–2 broods per year

Eggs: 3–6; pale blue without markings

Incubation: 11–12 days; female incubates

Fledging: 9–10 days; female and male feed the young

Migration: complete, to the Bahamas, Cuba, Mexico and Central America

Food: insects, seeds; will come to seed feeders

Compare: The female Indigo Bunting (p. 137) is very similar, but is more common, lacks the tan wing bars and it is lighter in color overall.

Stan's Notes: Summer resident and migrant. Returns to Louisiana and Mississippi by mid-April and leaves in October. It has expanded northward, and its overall populations have increased over the past 30–40 years. A bird of semi-open habitats such as overgrown fields, riversides, woodland edges and fencerows. Visits seed feeders, where it is often confused with female Indigo Buntings. Often seen twitching and spreading its tail. The first-year males show only some blue, obtaining the full complement of blue feathers in the second winter.

male
p. 27

female

YEAR-ROUND

Brown-headed Cowbird
Molothrus ater

Size: 7½" (19 cm)

Female: Dull brown with no obvious markings. Pointed, sharp, gray bill. Dark eyes.

Male: glossy black with a chocolate-brown head

Juvenile: similar to female but with dull-gray plumage and a streaked chest

Nest: no nest; lays eggs in the nests of other birds

Eggs: 5–7; white with brown markings

Incubation: 10–13 days; host bird incubates the eggs

Fledging: 10–11 days; host birds feed the young

Migration: non-migrator in Louisiana and Mississippi

Food: insects, seeds; will come to seed feeders

Compare: The female Red-winged Blackbird (p. 175) has white eyebrows and heavy streaking. The European Starling (p. 29) has speckles and a shorter tail. The pointed gray bill helps identify female Brown-headed Cowbird.

Stan's Notes: Cowbirds are members of the blackbird family. Known as brood parasites, Brown-headed Cowbirds are the only parasitic birds in Louisiana and Mississippi. Brood parasites lay their eggs in the nests of other birds, leaving the host birds to raise their young. Cowbirds are known to have laid their eggs in the nests of over 200 species of birds. While some birds reject cowbird eggs, most incubate them and raise the young, even to the exclusion of their own. Look for warblers and other birds feeding young birds twice their own size. Named "Cowbird" for its habit of following bison and cattle herds to feed on insects flushed up by the animals. Numbers may increase during winter, when migratory birds from the north join resident birds for the season.

WINTER

Cedar Waxwing
Bombycilla cedrorum

Size: 7½" (19 cm)

Male: Sleek-looking, gray-to-brown bird. Pointed crest, bandit-like mask and light-yellow belly. Bold-yellow tip of tail. Red wing tips look like they were dipped in red wax.

Female: same as male

Juvenile: grayish with a heavily streaked breast; lacks the sleek look, black mask and red wing tips

Nest: cup; female and male construct; 1 brood per year, occasionally 2

Eggs: 4–6; pale blue with brown markings

Incubation: 10–12 days; female incubates

Fledging: 14–18 days; female and male feed the young

Migration: complete, to Louisiana and Mississippi; moves around to find food

Food: cedar cones, fruit, insects

Compare: The female Northern Cardinal (p. 173) has a large red bill. Look for the red wing tips to identify the Cedar Waxwing.

Stan's Notes: The name is derived from its red, wax-like wing tips and preference for the small, berry-like cones of the cedar. Seen in flocks, moving around from area to area looking for berries. Nests in northern states. Arrives in Louisiana and Mississippi in fall and stays through winter. Wanders during winter, searching for food supplies. Spends most of its time at the tops of tall trees. Listen for the high-pitched "sreee" whistling sound it constantly makes while perched or in flight. Obtains the mask after the first year and red wing tips after the second year.

female

male
p. 25

Eastern Towhee

Pipilo erythrophthalmus

YEAR-ROUND
WINTER

Size: 7–8" (18–20 cm)

Female: Mostly light-brown bird. Rusty red-brown sides and a white belly. Long brown tail with a white tip. Short, stout, pointed bill and rich, red eyes. White wing patches flash in flight.

Male: similar to female but black instead of brown

Juvenile: light brown with a heavily streaked head, chest and belly; long dark tail with a white tip

Nest: cup; female builds; 2 broods per year

Eggs: 3–4; creamy white with brown markings

Incubation: 12–13 days; female incubates

Fledging: 10–12 days; male and female feed the young

Migration: partial to non-migrator in Louisiana and Mississippi; moves around to find food

Food: insects, seeds, fruit; visits ground feeders

Compare: The American Robin (p. 279) is larger, has a red breast and lacks the white belly. The female Rose-breasted Grosbeak (p. 165) has a heavily streaked breast and obvious white eyebrows.

Stan's Notes: Named for its distinctive "tow-hee" call (given by both sexes) but known mostly for its other characteristic call, which sounds like "drink-your-tea!" Will hop backward with both feet (bilateral scratching), raking up leaf litter to locate insects and seeds. The female broods, but male does the most feeding of young. In some southern coastal states, some have red eyes and others have white eyes. Red-eyed variety seen in Louisiana and Mississippi.

male
p. 59

female

Rose-breasted Grosbeak
Pheucticus ludovicianus

MIGRATION

Size: 7–8" (18–20 cm)

Female: Plump and heavily streaked. Large, obvious white eyebrows. Large ivory bill. Orange-to-yellow wing linings.

Male: black and white with a triangular rose patch in the center of the chest; rose wing linings

Juvenile: similar to female

Nest: cup; female and male construct; 1–2 broods per year

Eggs: 3–5; blue-green with brown markings

Incubation: 13–14 days; female and male incubate

Fledging: 9–12 days; female and male feed the young

Migration: complete, to Mexico, Central America and South America

Food: insects, seeds, fruit; comes to seed feeders

Compare: Looks like a large sparrow with bold white eyebrows and heavy streaking. The female Purple Finch (p. 145) has smaller eyebrows. The female House Finch (p. 131) lacks eyebrows.

Stan's Notes: Seen in small groups across Louisiana and Mississippi during spring and fall migration. Both sexes sing, but the male sings much louder and clearer. Sings a rich, robin-like song with a chip note in the tune. Often prefers to nest in mature deciduous forests. "Grosbeak" refers to the thick, strong bill, which is used to crush seeds. Males arrive at the breeding grounds a few days before females. Several individuals will come to seed feeders at the same time during migration. Both Louisiana and Mississippi are important stopovers, so feeders should be kept full during this time. Nests in northern states.

winter

breeding

Spotted Sandpiper
Actitis macularius

MIGRATION
WINTER

Size: 8" (20 cm)

Male: Olive-brown back with black spots on a white chest and belly. White line over eyes. Long, dull-yellow legs. Long bill. Winter plumage lacks spots on the chest and belly.

Female: same as male

Juvenile: similar to winter plumage, with a darker bill

Nest: ground; male builds; 2 broods per year

Eggs: 3–4; brownish with brown markings

Incubation: 20–24 days; male incubates

Fledging: 17–21 days; male feeds the young

Migration: complete, to Louisiana, Mississippi, other southern coastal states, Mexico, Central and South America

Food: aquatic insects

Compare: The Greater Yellowlegs (p. 193) is much larger. The Killdeer (p. 183) has 2 black neck bands. Look for the black spots on the chest and belly and the bobbing tail to help identify the breeding Spotted Sandpiper.

Stan's Notes: Seen along the shorelines of large ponds, lakes and rivers. One of the few shorebirds that will dive underwater when pursued. Able to fly straight up out of the water. Holds wings in a cup-like arc in flight, rarely lifting them above a horizontal plane. Walks as if delicately balanced. When standing, constantly bobs its tail. Female mates with multiple males and lays eggs in up to five nests. Male does all of the nest building, incubating and childcare without any help from the female. In winter plumage, it lacks the black spots on chest and belly.

winter
p. 269

breeding

Sanderling

Calidris alba

Size: 8" (20 cm)

Male: Breeding adult (April to August) has a rusty head, chest and back with white belly. Black legs and bill.

Female: same as male

Juvenile: spotty black on the head and back, a white belly, black legs and bill

Nest: ground; male builds; 1–2 broods per year

Eggs: 3–4; greenish olive with brown markings

Incubation: 24–30 days; male and female incubate

Fledging: 16–17 days; female and male feed the young

Migration: complete, to Louisiana, Mississippi, other southern coastal states, Mexico, and Central and South America

Food: insects

Compare: Spotted Sandpiper (p. 167) is the same size as Sanderling, but the breeding Spotted Sandpiper has black spots on its chest.

Stan's Notes: A migrant and winter shorebird, one of the most common shorebirds in Louisiana and Mississippi. Can be seen in groups on sandy beaches, running out with each retreating wave to feed. Look for a flash of white on the wings when it is in flight. Sometimes a female will mate with several males (polyandry), which results in males and the female incubating separate nests. Both sexes perform a distraction display if threatened. Nests on the Arctic tundra. Rests by standing on one leg (see inset) and tucking the other leg into its belly feathers. Often hops away on one leg, moving away from pedestrians on the beach. Surveys show a greater than 80 percent decline in numbers since the 1970s.

winter
p. 271

breeding

Dunlin

Calidris alpina

MIGRATION
WINTER

Size: 8–9" (20–23 cm)

Male: Breeding adult is distinctive with a rusty-red back, finely streaked chest and an obvious black patch on the belly. Stout bill, curving slightly downward at the tip. Black legs.

Female: slightly larger than male, with a longer bill

Juvenile: slightly rusty back with a spotty chest

Nest: ground; male and female construct; 1 brood per year

Eggs: 2–4; olive-buff or blue-green with red-brown markings

Incubation: 21–22 days; male incubates during the day, female incubates at night

Fledging: 19–21 days; male feeds the young, female often leaves before the young fledge

Migration: complete, to Louisiana and Mississippi, other southern coastal states, Mexico and Central America

Food: insects

Compare: Breeding Sanderling (p. 169) is similar in size, but look for the obvious black belly patch and curved bill of breeding Dunlin.

Stan's Notes: Breeding plumage more commonly seen in spring just before migrating north. Flights include heights of up to 100 feet (30 m) with brief gliding alternating with shallow flutters, and a rhythmic, repeating song. Huge flocks fly synchronously, with birds twisting and turning, flashing light and dark undersides. Males tend to fly farther south in winter than females. Doesn't nest in Louisiana and Mississippi.

male
p. 333

female

juvenile

Northern Cardinal

Cardinalis cardinalis

YEAR-ROUND

Size: 8–9" (20–23 cm)

Female: Buff-brown with red tinges on the crest and wings. Black mask and a large reddish bill.

Male: red with a large crest and bill and a black mask extending from the face to the throat

Juvenile: same as female but with a blackish-gray bill

Nest: cup; female builds; 2–3 broods per year

Eggs: 3–4; bluish white with brown markings

Incubation: 12–13 days; female and male incubate

Fledging: 9–10 days; female and male feed the young

Migration: non-migrator

Food: seeds, insects, fruit; comes to seed feeders

Compare: The Cedar Waxwing (p. 161) has a small dark bill. The juvenile Northern Cardinal (bottom inset) looks like the adult female but with a dark bill. Look for the reddish bill to identify the female Northern Cardinal.

Stan's Notes: A familiar backyard bird. Seen in a variety of habitats, including parks. Usually likes thick vegetation. One of the few species in which both females and males sing. Can be heard all year. Listen for its "whata-cheer-cheer-cheer" territorial call in spring. Watch for a male feeding a female during courtship. The male also feeds the young of the first brood while the female builds a second nest. Territorial in spring, fighting its own reflection in a window or other reflective surface. Non-territorial in winter, gathering in small flocks of up to 20 birds. Makes short flights from cover to cover, often landing on the ground. *Cardinalis* denotes importance, as represented by the red priestly garments of Catholic cardinals.

male
p. 31

female

Red-winged Blackbird
Agelaius phoeniceus

YEAR-ROUND

Size: 8½" (22 cm)

Female: Heavily streaked brown body. Pointed brown bill and white eyebrows.

Male: jet black with red-and-yellow shoulder patches (epaulets) and a pointed black bill

Juvenile: same as female

Nest: cup; female builds; 2–3 broods per year

Eggs: 3–4; bluish green with brown markings

Incubation: 10–12 days; female incubates

Fledging: 11–14 days; female and male feed the young

Migration: non-migrator in Louisiana and Mississippi

Food: seeds, insects; visits seed and suet feeders

Compare: The female Rose-breasted Grosbeak (p. 165) is plumper and has a thicker bill. The female Brown-headed Cowbird (p. 159) lacks streaks. Look for white eyebrows and heavy streaking to identify the female Red-winged.

Stan's Notes: One of the most widespread and numerous birds in Louisiana and Mississippi. In fall and winter, migrant and resident Red-wingeds gather in huge numbers (thousands) with other blackbirds to feed in agricultural fields, marshes, wetlands, and lakes and rivers. Flocks with as many as 10,000 birds have been reported. Males arrive before females and sing to defend their territory. The male repeats his call from the top of a cattail while showing off his red-and-yellow shoulder patches. The female chooses a mate and often builds her nest over shallow water in a thick stand of cattails. The male can be aggressive when defending the nest. Feeds mostly on seeds in spring and fall, and insects throughout the summer.

female

male

SUMMER

Common Nighthawk
Chordeiles minor

Size: 9" (23 cm)

Male: Camouflaged brown and white with a white chin. Distinctive white band across the wings and tail, seen only in flight.

Female: similar to male, with a tan chin; lacks a white tail band

Juvenile: similar to female

Nest: no nest; lays eggs on the ground, usually on rocks, or on rooftop; 1 brood per year

Eggs: 2; cream with lavender markings

Incubation: 19–20 days; female and male incubate

Fledging: 20–21 days; female and male feed the young

Migration: complete, to South America

Food: insects caught in the air

Compare: The Chimney Swift (p. 125) is much smaller. Look for the white chin, obvious white band on the wings and characteristic flap-flap-flap-glide pattern to help identify.

Stan's Notes: Usually only seen in flight at dusk or after sunset but not uncommon to see it sleeping on a branch during the day. A prolific insect eater and very noisy in flight, repeating a "peenting" call. Alternates slow wingbeats with bursts of quick wingbeats. In cities, prefers to nest on flat rooftops with gravel. City populations are on the decline as gravel rooftops are converted to other styles. Nests on the ground in the country. In spring, the male performs a showy mating ritual of a steep diving flight that ends with a loud popping noise. One of the first birds to migrate in fall. Seen in large flocks, all heading south.

in flight

male

juvenile

female

in-flight
juvenile

American Kestrel
Falco sparverius

YEAR-ROUND
WINTER

Size: 9–11" (23–28 cm); up to 2' wingspan

Male: Rust-brown back and tail. White breast with dark spots. Two vertical black lines on a white face. Blue-gray wings. Wide black band with a white edge on the tip of a rusty tail.

Female: similar to male but slightly larger, with rust-brown wings and dark bands on the tail

Juvenile: same as adult of the same sex

Nest: cavity; does not build a nest; 1 brood per year

Eggs: 4–5; white with brown markings

Incubation: 29–31 days; male and female incubate

Fledging: 30–31 days; female and male feed the young

Migration: partial to non-migrator in Louisiana and Mississippi; moves around to find food

Food: insects, small mammals and birds, reptiles

Compare: Similar to other falcons. Look for 2 vertical black stripes on its face. No other small bird of prey has a rusty back and tail.

Stan's Notes: An unusual raptor because the sexes look different (dimorphic). Due to its small size, this falcon was once called a Sparrow Hawk. Hovers near roads, then dives for prey. Watch for it to pump its tail after landing on a perch. Perches nearly upright. Eats many grasshoppers. Adapts quickly to a wooden nest box. Has pointed sweptback wings, seen in flight. Can be extremely vocal, giving a loud series of high-pitched calls. Ability to see ultraviolet (UV) light helps it locate mice and other prey by their urine, which glows bright yellow in UV light.

YEAR-ROUND

Northern Bobwhite
Colinus virginianus

Size: 10" (25 cm)

Male: Short, stocky and mostly brown with short gray tail. Prominent white eye stripe and white chin. Reddish-brown sides and belly, often with black lines and dots.

Female: similar to male, with buff-brown eye stripe and chin

Juvenile: smaller and duller than adults

Nest: ground; female and male construct; 1 brood per year

Eggs: 12–15; white to creamy without markings

Incubation: 23–24 days; female and male incubate

Fledging: 6–7 days; female and male feed the young

Migration: non-migrator

Food: insects, seeds, fruit; will come to ground feeders offering corn and millet

Compare: Mourning Dove (p. 189) is light brown and has a long pointed tail. Look for a small chicken-shaped bird that is usually seen in small groups.

Stan's Notes: Prefers shrubs, orchards, hedgerows and pastures. Moves around in small flocks of 20 birds (often family members), called a covey. The covey often rests together during the night, in a tight circle with tails together and heads facing outward, to watch for predators. Males and females perform distraction displays when nests or young are threatened. Nest is a depression in the ground lined with grass. Often pulls nearby vegetation over nest to help conceal it. Male gives a rising whistle, "bob-white," heard mainly in spring and summer. Also gives a single "hoy" call year-round.

Killdeer

Charadrius vociferus

YEAR-ROUND

Size: 11" (28 cm)

Male: Upland shorebird with 2 black bands around the neck, like a necklace. Brown back and white belly. Bright reddish-orange rump, visible in flight.

Female: same as male

Juvenile: similar to adults, with a single neck band

Nest: ground; male scrapes; 2 broods per year

Eggs: 3–5; tan with brown markings

Incubation: 24–28 days; male and female incubate

Fledging: 25 days; male and female lead their young to food

Migration: complete, to southern states, Mexico and Central America; non-migrator in Louisiana and Mississippi

Food: insects, worms, snails

Compare: The Spotted Sandpiper (p. 167) is found around water and lacks the 2 neck bands of the Killdeer.

Stan's Notes: Technically a shorebird but lives in dry habitats. Often found in vacant fields, driveways, wetland edges or along railroad tracks. The only shorebird that has two black neck bands. Known to fake a broken wing to draw intruders away from the nest; once the nest is safe, the parent will take flight. Nests are just a slight depression in a dry area and are often hard to see. Hatchlings look like miniature adults walking on stilts. Soon after hatching, the young follow their parents around and peck for insects. Gives a loud and distinctive "kill-deer" call. Migrates in small flocks.

Brown Thrasher
Toxostoma rufum

YEAR-ROUND

Size: 11" (28 cm)

Male: Rust-red with a long tail. Heavy streaking on the breast and belly. Two white wing bars. Long, curved bill and bright-yellow eyes.

Female: same as male

Juvenile: same as adults but with grayish eyes

Nest: cup; female and male build; 2 broods per year

Eggs: 4–5; pale blue with brown markings

Incubation: 11–14 days; female and male incubate

Fledging: 10–13 days; female and male feed the young

Migration: complete, to southern states; non-migrator in Louisiana and Mississippi

Food: insects, fruit

Compare: American Robin (p. 279) and Gray Catbird (p. 275) are similar in shape and slightly smaller in size, but the Brown Thrasher has a streaked chest, rusty color and yellow eyes.

Stan's Notes: A prodigious songster. Sings along forest edges and in suburban yards. Often found in thick shrubs, where it will sing deliberate musical phrases, repeating each twice. The male Brown Thrasher has the largest documented repertoire of all North American songbirds, with more than 1,100 types of songs. Builds nest low in dense shrubs, often in fencerows. Quickly flies or runs on the ground in and out of thick shrubs. A noisy feeder due to its habit of turning over leaves, small rocks and branches to find food. Populations increase during winter months with the influx of northern birds. This bird is more abundant in the central Great Plains than anywhere else in North America.

male

female

Northern Flicker

Colaptes auratus

YEAR-ROUND

Size: 12" (30 cm)

Male: Brown and black with a black mustache and black necklace. Red spot on the nape of the neck. Speckled chest. Large white rump patch, seen only when flying.

Female: same as male but without a black mustache

Juvenile: same as adult of the same sex

Nest: cavity; female and male excavate; 1 brood per year

Eggs: 5–8; white without markings

Incubation: 11–14 days; female and male incubate

Fledging: 25–28 days; female and male feed the young

Migration: non-migrator in Louisiana and Mississippi

Food: insects (especially ants and beetles); comes to suet feeders

Compare: The male Yellow-bellied Sapsucker (p. 61) has a red chin. The male Red-bellied Woodpecker (p. 67) has a red crown. Flickers are the only brown-backed woodpeckers in Louisiana and Mississippi.

Stan's Notes: This is the only woodpecker to regularly feed on the ground. Prefers ants and beetles and produces an antacid saliva that neutralizes the acidic defense of ants. The male often picks the nest site. Parents take up to 12 days to excavate the cavity. Can be attracted to your yard with a nest box stuffed with sawdust. Often reuses an old nest. Undulates deeply during flight, flashing yellow under its wings and tail and calling "wacka-wacka" loudly. Populations swell in winter with northern migrants.

Mourning Dove
Zenaida macroura

YEAR-ROUND

Size: 12" (30 cm)

Male: Smooth and fawn-colored. Gray patch on the head. Iridescent pink and greenish blue on the neck. Black spot behind and below the eyes. Black spots on the wings and tail. Pointed, wedged tail; white edges seen in flight.

Female: similar to male, but lacks the pink-and-green iridescent neck feathers

Juvenile: spotted and streaked plumage

Nest: platform; female and male build; 2 broods per year

Eggs: 2; white without markings

Incubation: 13–14 days; male incubates during the day, female incubates at night

Fledging: 12–14 days; female and male feed the young

Migration: non-migrator in Louisiana and Mississippi; will move around to find food

Food: seeds; will visit seed and ground feeders

Compare: Lacks the wide range of color combinations of the Rock Pigeon (p. 291). Lacks the black collar of the Eurasian Collared-Dove (p. 289).

Stan's Notes: Name comes from its mournful cooing. A ground feeder, bobbing its head as it walks. One of the few birds to drink without lifting its head, like the Rock Pigeon. The parents feed the young (squab) a regurgitated liquid called crop-milk during their first few days of life. Platform nest is so flimsy it often falls apart in a storm. During takeoff and in flight, wind rushes through the bird's wing feathers, creating a characteristic whistling sound.

winter

breeding

YEAR-ROUND

Pied-billed Grebe
Podilymbus podiceps

Size: 12–14" (30–36 cm)

Male: Small and brown with a black chin and fluffy white patch beneath the tail. Black ring around a thick, chicken-like, ivory bill. Winter bill is brown and unmarked.

Female: same as male

Juvenile: paler than adults, with white spots and a gray chest, belly and bill

Nest: floating platform; female and male build; 1 brood per year

Eggs: 5–7; bluish white without markings

Incubation: 22–24 days; female and male incubate

Fledging: 45–60 days; female and male feed the young

Migration: complete, to southern states, Mexico and Central America; non-migrator in Louisiana and Mississippi

Food: crayfish, aquatic insects, fish

Compare: Look for a puffy white patch under the tail and thick, chicken-like bill to help identify.

Stan's Notes: A common resident water bird, often seen diving for food. When disturbed, it slowly sinks like a submarine, quickly compressing its feathers, forcing the air out. Was called Hell-diver due to the length of time it can stay submerged. Able to surface far from where it went under. Well suited to life on water, with short wings, lobed toes, and legs set close to the rear of its body. Swims easily but moves awkwardly on land. Very sensitive to pollution. Builds nest on a floating mat in water. "Grebe" may originate from the Breton word *krib*, meaning "crest," referring to the crested head plumes of many grebes, especially during breeding season.

Greater Yellowlegs
Tringa melanoleuca

MIGRATION
WINTER

Size: 13–15" (33–38 cm)

Male: Tall with a bulbous head and a long, thin, slightly upturned bill. Gray streaking on the chest. White belly. Long yellow legs.

Female: same as male

Juvenile: same as adults

Nest: ground; female builds; 1 brood per year

Eggs: 3–4; off-white with brown markings

Incubation: 22–23 days; female and male incubate

Fledging: 18–20 days; male and female feed the young

Migration: complete, to Louisiana, Mississippi, other southern states, Mexico and Central and South America

Food: small fish, aquatic insects

Compare: Breeding Willet (p. 195) is slightly larger, with a shorter neck and larger head, and it lacks bright-yellow legs. The breeding Spotted Sandpiper (p. 167) has spots on its chest. Look for the long yellow legs and long bill to identify the Greater Yellowlegs.

Stan's Notes: A common shorebird that can be identified by the slightly upturned bill and long yellow legs, which enable it to wade in deep water. Often seen resting on one leg. Rushes forward through the water to feed, plowing its bill or swinging it from side to side, catching small fish and insects. A skittish bird, it is quick to give an alarm call, causing flocks to take flight. Typically moves into the water before taking flight. Gives a variety of "flight" notes at takeoff. Nests on the ground close to water on the northern tundra of Labrador and Newfoundland.

breeding

winter
p. 293

displaying

Willet
Catoptrophorus semipalmatus

YEAR-ROUND

Size: 14–16" (36–40 cm)

Male: Brown breeding plumage with a white belly. Brown bill and legs. Distinctive black-and-white wing lining pattern, seen in flight or during display.

Female: same as male

Juvenile: similar to breeding adult, more tan in color

Nest: ground; female builds; 1 brood per year

Eggs: 3–5; olive-green with dark markings

Incubation: 24–28 days; male and female incubate

Fledging: 1–2 days; female and male feed young

Migration: non-migrator in southern parts of Louisiana and Mississippi

Food: insects, small fish, crabs, worms, clams

Compare: Greater Yellowlegs (p. 193) is slightly smaller and has a smaller head, longer neck and yellow legs. Killdeer (p. 183) has 2 black bands on the neck.

Stan's Notes: A year-round resident in Louisiana and Mississippi. One of the few shorebirds that is seen away from water, sometimes standing on fence posts. It appears a rich, warm brown during the breeding season and rather plain gray during the winter, but it always has a striking black-and-white wing pattern when seen in flight. Uses its black-and-white wing patches to display to its mate. Named after the "pill-will-willet" call it gives during the breeding season. Gives a "kip-kip-kip" alarm call when it takes flight. Nests along the Gulf and East Coasts, in some western states and Canada.

male
p. 39

female

Boat-tailed Grackle

Quiscalus major

YEAR-ROUND

Size: 13–15" (33–38 cm), female
15–17" (38–43 cm), male

Female: A golden-brown chest and head. Nearly black wings and tail. Female is non-iridescent.

Male: iridescent blue-black bird with a very long tail and dark eyes

Juvenile: similar to female

Nest: cup; female builds; 2 broods per year

Eggs: 2–4; pale greenish blue with brown marks

Incubation: 13–15 days; female incubates

Fledging: 12–15 days; female feeds the young

Migration: non-migrator; moves around to find food

Food: insects, berries, seeds, fish; visits feeders

Compare: Fairly distinctive. Not confused with many other birds.

Stan's Notes: A noisy bird of coastal saltwater and inland marshes, giving several harsh, high-pitched calls and several squeaks. Eats a wide variety of foods from grains to fish. Sometimes seen picking insects off the backs of cattle. Will also visit bird feeders. Makes a cup nest with mud or cow dung and grass. Nests in small colonies. Most nesting occurs from February through July and occasionally again from October to December. Much less widespread in Louisiana and Mississippi then the Common Grackle and Great-tailed Grackle. Boat-taileds on the Gulf Coast have dark eyes, while Atlantic Coast birds have bright-yellow eyes.

male
p. 41

female

YEAR-ROUND

Great-tailed Grackle
Quiscalus mexicanus

Size: 15" (38 cm), female
18" (45 cm), male

Female: An overall brown bird with a gray-to-brown belly. Light-brown-to-white eyes, eyebrows, throat and upper portion of chest.

Male: all-black bird with iridescent purple sheen on the head and back, exceptionally long tail, bright-yellow eyes

Juvenile: similar to female

Nest: cup; female builds; 1–2 broods per year

Eggs: 3–5; greenish blue with brown markings

Incubation: 12–14 days; female incubates

Fledging: 21–23 days; female feeds the young

Migration: non-migrator to partial in Louisiana; moves around to find food

Food: insects, fruit, seeds; comes to seed feeders

Compare: The female Boat-tailed Grackle (p. 197) is similar but is found along the coast. The female Brown-headed Cowbird (p. 159) is much smaller than the Great-tailed female.

Stan's Notes: This is our largest grackle. It was once considered a subspecies of the Boat-tailed Grackle, which occurs along the Gulf Coast and in the southern part of Louisiana and Mississippi. Prefers to nest close to water in an open habitat. A colony nester. Males do not participate in nest building, incubation or raising young. Males rarely fight; females squabble over nest sites and materials. Several females mate with one male. The species is expanding northward, moving into northern states. Western populations tend to be larger than eastern. Song varies from population to population.

199

male
p. 75

female

Bufflehead
Bucephala albeola

WINTER

Size: 13–15" (33–38 cm)

Female: Brownish-gray duck with a dark-brown head. White patch on cheek, just behind the eyes.

Male: striking black-and-white duck with a large bonnet-like white patch on the back of head; head shines greenish purple in sunlight

Juvenile: similar to female

Nest: cavity; female lines an old woodpecker cavity; 1 brood per year

Eggs: 8–10; ivory-to-olive without markings

Incubation: 29–31 days; female incubates

Fledging: 50–55 days; female leads the young to food

Migration: complete to Louisiana, Mississippi, other southern states and Mexico

Food: aquatic insects, crustaceans, mollusks

Compare: Female Lesser Scaup (p. 209) is slightly larger and has a white patch at the base of the bill. Look for the white cheek patch to help identify the female Bufflehead.

Stan's Notes: A small, common diving duck, almost always seen in small groups or with other duck species during migration and winter on rivers, ponds and lakes across Louisiana and Mississippi. Nests in vacant woodpecker holes. When cavities in trees are scarce, known to use a burrow in an earthen bank or will use a nest box. Lines the cavity with fluffy down feathers. Unlike other ducks, the young stay in the nest for up to two days before they venture out with their mothers. The female is very territorial and remains with the same mate for many years. Does not breed in Louisiana and Mississippi.

soaring

Red-shouldered Hawk

Buteo lineatus

YEAR-ROUND

Size: 15–19" (38–48 cm); up to 3½' wingspan

Male: Reddish (cinnamon) head, shoulders, breast and belly. Wings and back are dark brown with white spots. Long tail with thin white bands and wide black bands. Obvious red wing linings, seen in flight.

Female: same as male

Juvenile: similar to adults but lacks the cinnamon color; white chest with dark spots

Nest: platform; female and male build; 1 brood per year

Eggs: 2–4; white with dark markings

Incubation: 27–29 days; female and male incubate

Fledging: 39–45 days; female and male feed the young

Migration: non-migrator to partial; moves around to find food

Food: reptiles, amphibians, large insects, birds

Compare: The Red-tailed Hawk (p. 227) has a white chest. The Sharp-shinned Hawk (p. 285) is smaller and lacks the reddish head and belly of the Red-shouldered Hawk.

Stan's Notes: A common woodland hawk throughout Louisiana and Mississippi, seen in backyards. Likes to hunt at forest edges, spotting snakes, frogs, insects, occasional small birds and other prey as it perches. Often flaps with an alternating gliding pattern. Very vocal with a distinct scream. Breeds when it reaches 2–3 years. Remains in the same territory for many years. Starts constructing its nest in February. Young leave the nest by June.

Green-winged Teal
Anas crecca

WINTER

Size:	14–15" (36–38 cm)
Male:	Gray body. Chestnut head with a dark-green patch outlined with white from the eyes to the nape of the neck. Green patch on the wings (speculum), seen in flight. Butter-yellow tail.
Female:	light brown with black spots and a green speculum; small bill
Juvenile:	same as female
Nest:	ground; female builds; 1 brood per year
Eggs:	8–10; creamy white without markings
Incubation:	21–23 days; female incubates
Fledging:	32–34 days; female teaches the young to feed
Migration:	complete, to Louisiana and Mississippi; moves around to find food
Food:	aquatic plants and insects
Compare:	The female Blue-winged Teal (p. 207) is similar in size, but it has slight white at the base of its bill. Look for the chestnut head with a dark-green patch on each side to identify the male Green-winged Teal.

Stan's Notes: One of the smallest dabbling ducks. Tips forward in water to feed off the bottom of shallow ponds. This behavior makes it vulnerable to ingesting spent lead shot, which can cause death. It walks well on land and will also feed in flooded fields and woodlands. Known for its fast and agile flight. Groups wheel and spin through the air in tight formation. The green wing patches are most obvious during flight.

male

female

Blue-winged Teal
Spatula discors

YEAR-ROUND
SUMMER
MIGRATION
WINTER

Size: 15–16" (38–41 cm)

Male: Small, plain-looking brown duck with black speckles and a large, crescent-shaped white mark at the base of the bill. Gray head. Black tail with a small white patch. Blue wing patch (speculum), best seen in flight.

Female: duller than male, with only slight white at the base of the bill; lacks a crescent mark on the face and a white patch on the tail

Juvenile: same as female

Nest: ground; female builds; 1 brood per year

Eggs: 8–11; creamy white

Incubation: 23–27 days; female incubates

Fledging: 35–44 days; female feeds the young

Migration: complete to non-migrator in Louisiana and Mississippi

Food: aquatic plants, seeds, aquatic insects

Compare: Female Green-winged Teal (p. 205) lacks white near its bill. The female Mallard (p. 219) is larger and has an orange-and-black bill. The female Wood Duck (p. 215) is larger and has a crest. Look for the white facial mark to identify the male Blue-winged.

Stan's Notes: One of the smallest ducks in North America and one of the longest-distance migrating ducks, with widespread nesting as far north as Alaska. One of the most widespread and abundant winter ducks in Louisiana and Mississippi. Arrives in August; leaves in April and May. Female performs a distraction display to protect nest and young. Male leaves female near the end of incubation.

male
p. 75

female

Lesser Scaup
Aythya affinis

WINTER

Size: 16–17" (40–43 cm)

Female: Overall brown duck with dull-white patch at the base of a light gray bill. Yellow eyes.

Male: white and gray; the chest and head appear nearly black but the head looks purple with green highlights in direct sun; yellow eyes

Juvenile: same as female

Nest: ground; female builds; 1 brood per year

Eggs: 8–14; olive-buff without markings

Incubation: 22–28 days; female incubates

Fledging: 45–50 days; female teaches the young to feed

Migration: complete, to Louisiana and Mississippi, other southern states, Mexico, Central America

Food: aquatic plants and insects

Compare: Female Ring-necked Duck (p. 213) is a similar size and has a white ring around the bill. Male Blue-winged Teal (p. 207) is slightly smaller and has a crescent-shaped white mark at the base of bill. The female Wood Duck (p. 215) is larger with white around eyes.

Stan's Notes: Often in large flocks on lakes, ponds and sewage lagoons in winter. Prefers fresh water, but can be seen along the coast. Completely submerges itself to feed on the bottom of lakes (unlike dabbling ducks, which only tip forward to reach bottom). Note the bold white stripe under wings when in flight. Male leaves female when she starts incubating eggs. Quantity of eggs (clutch size) increases with age of female. Interesting baby-sitting arrangement in which groups of young (crèches) are tended by 1–3 adult females.

male
p. 83

female

Hooded Merganser
Lophodytes cucullatus

WINTER

Size: 16–19" (41–48 cm)

Female: Sleek brown-and-rust bird with a red head. Ragged "hair" on the back of the head. Long, thin, brown bill.

Male: black back, rust-brown sides, long black bill; raises crest "hood" to display a white patch

Juvenile: similar to female

Nest: cavity; female lines an old woodpecker cavity or a nest box near water; 1 brood per year

Eggs: 10–12; white without markings

Incubation: 32–33 days; female incubates

Fledging: 71 days; female feeds the young

Migration: complete, to Louisiana, Mississippi and other southern states

Food: small fish, aquatic insects, crustaceans (especially crayfish)

Compare: Female Lesser Scaup (p. 209) is smaller and has a dull-white patch at the base of its bill.

Stan's Notes: A small diving duck, found in shallow ponds, sloughs, lakes and rivers. The male Hoodie voluntarily raises and lowers its crest to show off the large white head patch (hood). Rarely found away from wooded areas, where it nests in natural cavities or nest boxes. Usually in small groups. Quick, low flight across the water, with fast wingbeats. Male has a deep, rolling call. Female gives a hoarse quack. Nests in wooded areas. Female will lay some eggs in the nests of other mergansers or Wood Ducks (egg dumping), resulting in 20–25 eggs in some nests. Rarely, she shares a nest, sitting with a Wood Duck. When young hatch, there is often a mixture of Hoodies and Wood Ducks.

male
p. 81

female

Ring-necked Duck
Aythya collaris

WINTER

Size: 16–19" (41–48 cm)

Female: Brown with a darker brown back and crown and lighter-brown sides. Gray face. White eye-ring with a white line behind the eye. White ring around the bill. Peaked head.

Male: black head, chest and back; gray-to-white sides; blue bill with a bold white ring and a thinner ring at the base; peaked head

Juvenile: similar to female

Nest: ground; female builds; 1 brood per year

Eggs: 8–10; olive to brown without markings

Incubation: 26–27 days; female incubates

Fledging: 49–56 days; female teaches the young to feed

Migration: complete, to Louisiana and Mississippi, other southern states, Mexico and Central America

Food: aquatic plants and insects

Compare: The female Lesser Scaup (p. 209) is similar in size. Look for the white ring around the bill to help identify the female Ring-necked Duck.

Stan's Notes: A common winter resident in Louisiana and Mississippi. Often seen in larger freshwater lakes, usually in small flocks or just pairs. A diving duck, watch for it to dive underwater to forage for food. Springs up off the water to take flight. Has a distinctive tall, peaked head with a sloped forehead. Flattens its crown when diving. Named "Ring-necked" for its cinnamon collar, which is nearly impossible to see in the field. Also called Ring-billed Duck due to the white ring on its bill.

male
p. 315

female

Wood Duck

Aix sponsa

YEAR-ROUND

Size: 17–20" (43–51 cm)

Female: Small brown dabbling duck. Bright-white eye-ring and a not-so-obvious crest. Blue patch on wings (speculum), often hidden.

Male: highly ornamented, with a mostly green head and crest patterned with black and white; rusty chest, white belly and red eyes

Juvenile: similar to female

Nest: cavity; female lines an old woodpecker cavity or a nest box in a tree; 1 brood per year

Eggs: 10–15; creamy white without markings

Incubation: 28–36 days; female incubates

Fledging: 56–68 days; female teaches the young to feed

Migration: non-migrator in Louisiana and Mississippi

Food: aquatic insects, plants, seeds

Compare: The female Mallard (p. 219) and female Blue-winged Teal (p. 207) lack the eye-ring and crest. The female Northern Shoveler (p. 221) is larger and has a larger spoon-shaped bill.

Stan's Notes: A common duck of quiet, shallow backwater ponds. Nearly went extinct around 1900 due to overhunting, but it's doing well now. Nests in a tree cavity or a nest box in a tree. Seen flying in forests or perching on high branches. Female takes off with a loud, squealing call and enters the nest cavity from full flight. Lays some eggs in a neighboring nest (egg dumping), resulting in more than 20 eggs in some clutches. Hatchlings stay in the nest for 24 hours, then jump from as high as 60 feet (18 m) to the ground or water to follow their mother. They never return to the nest. Populations increase in winter with northern migrants.

American Wigeon
Anas americana

WINTER

Size: 18–20" (48 cm)

Male: Brown duck with a rounded head and obvious white cap. Deep-green patch starting behind the eyes and streaking down the neck. Long pointed tail. Short, black-tipped grayish bill. White belly and wing linings, seen in flight. Non-breeding lacks white cap and green patch.

Female: light brown with a pale gray head, a short, black-tipped grayish bill, green wing patch (speculum) and dark eye spot; white belly and wing linings, seen in flight

Juvenile: similar to female

Nest: ground; female builds; 1 brood per year

Eggs: 7–12; white without markings

Incubation: 23–25 days; female incubates

Fledging: 37–48 days; female teaches the young to feed

Migration: complete, to Louisiana, Mississippi, other southern states and Mexico

Food: aquatic plants, seeds

Compare: Male American Wigeon is easily identified by the white cap and black-tipped grayish bill. Look for the black-tipped grayish bill and green wing patch to help identify the female American Wigeon.

Stan's Notes: Often in small flocks or with other ducks. Male stays with the female only during the first week of incubation. Female raises the young. If threatened, female feigns injury while the young run and hide. Conceals upland nest in tall vegetation.

217

male
p. 317

female

WINTER

Mallard
Anas platyrhynchos

Size: 19–21" (48–53 cm)

Female: Brown duck with a blue-and-white wing mark (speculum). Orange-and-black bill.

Male: large green head, white necklace, rust-brown or chestnut chest, combination of gray-and-white sides, yellow bill, orange legs and feet

Juvenile: same as female but with a yellow bill

Nest: ground; female builds; 1 brood per year

Eggs: 7–10; greenish to whitish, unmarked

Incubation: 26–30 days; female incubates

Fledging: 42–52 days; female leads the young to food

Migration: complete, to Louisiana and Mississippi, other southern states

Food: seeds, plants, aquatic insects; will come to ground feeders offering corn

Compare: The Mottled Duck (p. 233) is similar, but has a yellow bill. The female Wood Duck (p. 215) has a white eye-ring. The female Blue-winged Teal (p. 207) is smaller than the female Mallard. Female Northern Shoveler (p. 221) has a large spoon-shaped bill.

Stan's Notes: A familiar dabbling duck of lakes and ponds. Also found in rivers, streams and some backyards. Tips forward to feed on vegetation on the bottom of shallow water. The name "Mallard" comes from the Latin word *masculus*, meaning "male," referring to the male's habit of taking no part in raising the young. Female and male have white underwings and white tails, but only the male has black central tail feathers that curl upward. The female gives a classic quack. Returns to its birthplace each year.

male
p. 319

female

Northern Shoveler
Anas clypeata

WINTER

Size: 19–21" (48–53 cm)

Female: Medium-size brown duck speckled with black. Green patch on the wings (speculum). Extraordinarily large, spoon-shaped bill.

Male: iridescent green head, rusty sides, white chest, large spoon-shaped bill

Juvenile: same as female

Nest: ground; female builds; 1 brood per year

Eggs: 9–12; olive without markings

Incubation: 22–25 days; female incubates

Fledging: 30–60 days; female leads the young to food

Migration: complete, to Louisiana, Mississippi, other southern states, Mexico and Central America

Food: aquatic insects, plants

Compare: The female Wood Duck (p. 215) is smaller and has a white eye-ring. The female Mallard (p. 219) lacks the large bill. Look for the spoon-shaped bill to identify the Shoveler.

Stan's Notes: One of several species of shovelers. Called "Shoveler" due to the peculiar shovel-like shape of its bill. Given the common name "Northern" because it is the only species of these ducks in North America. Seen in shallow wetlands, ponds and small lakes in flocks of 5–10 birds. Flocks fly in tight formation. Swims low in water, pointing its large bill toward the water as if it's too heavy to lift. Usually swims in tight circles while feeding. Feeds mainly by filtering tiny aquatic insects and plants from the surface of the water with its bill. Female gathers plant material and forms it into a nest a short distance from the water. A winter resident in Louisiana and Mississippi that arrives in September and leaves in April.

male
p. 335

female

Canvasback

Aythya valisineria

WINTER

Size: 20½" (52 cm)

Female: Brown head, neck and chest. Light-gray-to-brown sides. Long sloping forehead that transitions into a long dark bill.

Male: deep-red head and neck, sloping forehead, long black bill, gray-and-white sides and back, black chest and tail

Juvenile: similar to female

Nest: ground; female builds; 1 brood per year

Eggs: 7–9; pale white to gray without markings

Incubation: 24–29 days; female incubates

Fledging: 56–67 days; female leads young to food

Migration: complete, to Louisiana, Mississippi, other southern states and Mexico

Food: aquatic insects, small clams

Compare: Female Lesser Scaup (p. 209) is smaller, has a white marking at the base of its bill and lacks the sloping forehead and long dark bill of the female Canvasback.

Stan's Notes: A large inland duck of freshwater lakes, rivers and ponds. Populations declined dramatically in the 1960–80s due to marsh drainage for agriculture. Females return to their birthplace (philopatric) while males disperse to new areas. Will mate during migration or on the breeding grounds. A courting male gives a soft cooing call when displaying and during aerial chases. Male leaves the female after incubation starts. Female takes a new mate every year. Female feeds very little during incubation and will lose up to 70 percent of fat reserves during that time.

male
p. 299

female

soaring

Northern Harrier
Circus hudsonius

WINTER

Size: 18–22" (45–56 cm); up to 4' wingspan

Female: Slender, low-flying hawk with a dark-brown back and brown streaking on the chest and belly. Large white rump patch. Thin black tail bands and black wing tips. Yellow eyes.

Male: silver-gray with a large white rump patch and white belly; faint, thin bands across tail; black wing tips; yellow eyes

Juvenile: similar to female, with an orange breast

Nest: ground, sometimes low in a shrub; female and male construct; 1 brood per year

Eggs: 4–8; bluish white without markings

Incubation: 31–32 days; female incubates

Fledging: 30–35 days; male and female feed the young

Migration: complete, to Louisiana, Mississippi, other southern states, Mexico and Central America

Food: mice, snakes, insects, small birds

Compare: Slimmer than the Red-tailed Hawk (p. 227). Look for the characteristic low gliding and the black tail bands to identify the female Harrier.

Stan's Notes: One of the easiest of hawks to identify. Glides just above the ground, following the contours of the land while searching for prey. Holds its wings just above horizontal, tilting back and forth in the wind, similar to the Turkey Vulture. Formerly called Marsh Hawk due to its habit of hunting over marshes. Feeds and nests on the ground. Will also preen and rest on the ground. Unlike other hawks, mainly uses its hearing to find prey, followed by its sight. At any age, it has a distinctive owl-like face disk.

soaring

juvenile
soaring

juvenile

Red-tailed Hawk

Buteo jamaicensis

YEAR-ROUND

Size: 19–23" (48–58 cm); up to 4½' wingspan

Male: Variety of colorations, from chocolate brown to nearly all white. Often brown with a white breast and brown belly band. Rust-red tail. Underside of wing is white with a small dark patch on the leading edge near the shoulder.

Female: same as male but slightly larger

Juvenile: similar to adults, with a speckled breast and light eyes; lacks a red tail

Nest: platform; male and female build; 1 brood per year

Eggs: 2–3; white without markings or sometimes marked with brown

Incubation: 30–35 days; female and male incubate

Fledging: 45–46 days; male and female feed the young

Migration: non-migrator; moves around in winter to find food

Food: small and medium-size animals, large birds, snakes, fish, insects, bats, carrion

Compare: Red-shouldered Hawk (p. 203) and Sharp-shinned Hawk (p. 285) are much smaller.

Stan's Notes: Common in open country and cities. Seen perching on fences, freeway lampposts and trees. Look for it circling above open fields and roadsides, searching for prey. Gives a high-pitched scream that trails off. Often builds a large stick nest in large trees along roads. Lines nest with finer material, like evergreen needles. Returns to the same nest site each year. The red tail develops in the second year and is best seen from above. Populations increase in winter with the arrival of northern birds.

Barred Owl

Strix varia

YEAR-ROUND

Size: 20–24" (51–61 cm); up to 3½' wingspan

Male: Chunky brown-and-gray owl. Dark horizontal barring on upper chest. Vertical streaks on lower chest and belly. A large head and dark-brown eyes. Yellow bill and feet.

Female: same as male but slightly larger

Juvenile: light gray with a black face

Nest: cavity; does not add any nesting material; 1 brood per year

Eggs: 2–3; white without markings

Incubation: 28–33 days; female incubates

Fledging: 42–44 days; female and male feed the young

Migration: non-migrator

Food: mice, rabbits and other animals; small birds; fish; reptiles; amphibians

Compare: Great Horned Owl (p. 231) has "horns," and the much smaller Eastern Screech-Owl (p. 273) has ears, both of which Barred Owl lacks. Look for a stocky owl with a large head and dark-brown eyes to identify the Barred Owl.

Stan's Notes: A very common owl in the state. Prefers deciduous, dense woodlands with sparse undergrowth, but it can be attracted to your yard with a simple nest box that has a large entrance hole. Often seen hunting during the day. Perches and watches for mice, birds and other prey. Hovers over water and reaches down to grab fish. After fledging, the young stay with their parents for up to four months. Often sounds like a dog barking just before calling six to eight hoots, sounding like "who-who-who-cooks-for-you."

Great Horned Owl
Bubo virginianus

YEAR-ROUND

Size: 21–25" (53–64 cm); up to 4' wingspan

Male: Robust brown "horned" owl. Bright-yellow eyes and a V-shaped white throat resembling a necklace. Horizontal barring on the chest.

Female: same as male but slightly larger

Juvenile: similar to adults but lacks ear tufts

Nest: no nest; takes over the nest of a crow, hawk or Great Blue Heron or uses a partial cavity, stump or broken tree; 1 brood per year

Eggs: 2–3; white without markings

Incubation: 26–30 days; female incubates

Fledging: 30–35 days; male and female feed the young

Migration: non-migrator

Food: mammals, birds (ducks), snakes, insects

Compare: The Barred Owl (p. 229) has dark eyes and no "horns." The Eastern Screech-Owl (p. 273) is extremely tiny. Look for bright-yellow eyes and feather "horns" on the head to help identify the Great Horned Owl.

Stan's Notes: The largest owl in Louisiana and Mississippi and one of the earliest nesting birds in the state, laying eggs in January and February. Able to hunt in complete darkness due to its excellent hearing. The "horns," or "ears," are tufts of feathers and have nothing to do with hearing. Cannot turn its head all the way around. Wing feathers are ragged on the ends, resulting in silent flight. Eyelids close from the top down, like ours. Fearless, it is one of the few animals that will kill skunks and porcupines. Given that, it is also called the Flying Tiger. Call sounds like "hoo-hoo-hoo-hoooo."

female

in
flight

Mottled Duck

Anas fulvigula

YEAR-ROUND

Size: 21–23" (53–58 cm)

Male: All-brown duck with a light-tan neck and head. Bright-yellow bill without markings. Wing patch (speculum) is blue (sometimes green) and outlined with black.

Female: same as male

Juvenile: same as adult

Nest: ground; female builds; 1 brood per year

Eggs: 5–10; off-white without markings

Incubation: 25–27 days; female incubates

Fledging: 60–70 days; female shows young what to eat

Migration: non-migrator in the southern third of Louisiana and a small area of south central Mississippi

Food: aquatic insects, crayfish, snails, grass, seeds

Compare: Female Mallard (p. 219) has an orange bill with black markings, and its speculum is outlined with white. Look for the female Mottled Duck's black-outlined speculum.

Stan's Notes: A year-round resident, found in southern parts of Louisiana and Mississippi. It is most densely populated along coastal Louisiana and south-central areas of Mississippi in both freshwater and saltwater marshes. Feeds on more insects and crayfish than the Mallard. Will breed with Mallards, producing hybrids. Mated pairs will stay together all year, unlike Mallards. Young will scatter if mother feels threatened and gives them an alarm call.

breeding

winter

Glossy Ibis
Plegadis falcinellus

YEAR-ROUND

Size: 22–24" (56–60 cm); up to 3' wingspan

Male: Breeding male has a chestnut-brown head and neck. Iridescent green-and-blue wings and tail. Appears to be all dark brown from a distance. Very long, down-curved bill with blue facial skin near base. Winter is overall dark with a speckled head.

Female: same as male

Juvenile: same as adult, but lacks iridescent coloring

Nest: platform; female and male build; 1 brood per year

Eggs: 2–4; light blue without markings

Incubation: 20–21 days; female and male incubate

Fledging: 28–32 days; female and male feed the young

Migration: non-migrator in southern parts of Louisiana and Mississippi

Food: aquatic insects, crustaceans

Compare: One of two ibis species in Louisiana and Mississippi; the long, down-curved bill helps identify them. White Ibis (p. 353) is all white and not confused with the brown Glossy Ibis.

Stan's Notes: Seems to be on the increase in Louisiana and Mississippi. Prefers fresh water over salt water, with crayfish a big part of the diet. From a distance the bird appears dark brown or nearly black, but when seen up close or through binoculars, its iridescent green and bluish-purple colors are amazing. Its long down-curved bill helps identify it in flight. Often seen flying in groups of 30 or more. Nests in large colonies with other wading birds.

Wild Turkey

Meleagris gallopavo

YEAR-ROUND

Size: 36–48" (91–122 cm)

Male: Large brown-and-bronze bird with a naked blue-and-red head. Long, straight, black beard in the center of the chest. Tail spreads open like a fan. Spurs on legs.

Female: thinner and less striking than the male; often lacks a breast beard

Juvenile: same as adult of the same sex

Nest: ground; female builds; 1 brood per year

Eggs: 10–12; buff-white with dull-brown markings

Incubation: 27–28 days; female incubates

Fledging: 6–10 days; female leads the young to food

Migration: non-migrator; moves around to find food

Food: insects, seeds, fruit

Compare: This bird is quite distinctive and unlikely to be confused with others.

Stan's Notes: The largest native game bird in Louisiana and Mississippi, and the species from which the domestic turkey was bred. Almost became our national bird, losing to the Bald Eagle by a single vote. A strong flier that can approach 60 miles (97 km) per hour. Can fly straight up, then away. Eyesight is three times better than ours. Hearing is also excellent; can hear competing males up to a mile away. Male has a "harem" of up to 20 females. Female scrapes out a shallow depression for nesting and pads it with soft leaves. Males are known as toms, females are hens, and young are poults. Roosts in trees at night. Eliminated from many of the southern states by the turn of the 20th century due to market hunting and loss of habitat, and reintroduced. Populations are now stable.

juvenile

breeding

non-breeding

chick-feeding
adult

Brown Pelican

Pelecanus occidentalis

YEAR-ROUND

Size: 46–50" (117–127 cm); up to 7' wingspan

Male: Gray-brown body, black belly, exceptionally long gray bill. Breeding adult has a white or yellow head with dark chestnut hind neck. Adult that is feeding chicks (chick-feeding adult) has a speckled white head. A non-breeding adult has a white head and neck.

Female: similar to male

Juvenile: brown with white breast and belly; does not acquire adult plumage until the third year

Nest: ground; female and male build; 1 brood per year

Eggs: 2–4; white without markings

Incubation: 28–30 days; female and male incubate

Fledging: 71–86 days; female and male feed the young

Migration: non-migrator in southern parts of Louisiana and Mississippi

Food: fish

Compare: A large, unmistakable bird in Louisiana and Mississippi.

Stan's Notes: A coastal bird of Louisiana and Mississippi and recently removed from the endangered species list. Suffering from eggshell thinning in the 1970s due to DDT and other pesticides, it is now reestablishing along the East, Gulf and West Coasts. Captures fish by diving headfirst into the ocean, opening its large bill and "netting" fish with its gular pouch (an expandable membrane beneath its lower jaw). Often seen sitting on posts around marinas. Nests in large colonies. Doesn't breed before the age of 3, when it obtains its breeding plumage.

WINTER

Ruby-crowned Kinglet
Regulus calendula

Size: 4" (10 cm)

Male: Small, teardrop-shaped green-to-gray bird. Two white wing bars and a white eye-ring. Hidden ruby crown.

Female: same as male, but lacks a ruby crown

Juvenile: same as female

Nest: pendulous; female builds; 1 brood per year

Eggs: 4–5; white with brown markings

Incubation: 11–12 days; female incubates

Fledging: 11–12 days; female and male feed the young

Migration: complete, to Louisiana, Mississippi, other southern states and Mexico

Food: insects, berries

Compare: The female American Goldfinch (p. 359) shares the drab olive plumage and unmarked chest, but it is larger. Look for the white eye-ring to identify the Ruby-crowned Kinglet.

Stan's Notes: This is one of the smallest birds in Louisiana and Mississippi. Most commonly seen in the winter. Look for it flitting around thick shrubs low to the ground. It takes a quick eye to see the ruby crown, which the male flashes when he is excited. The female weaves an unusually intricate nest and fastens colorful lichens and mosses to the exterior with spiderwebs. Often builds the nest high in a mature tree, where it hangs from a branch that has overlapping leaves. Doesn't nest in Louisiana and Mississippi. Sings a distinctive song that starts out soft and ends loud and on a higher note. "Kinglet" originates from the word *king*, referring to the male's red crown, and the diminutive suffix *let*, meaning "small."

Red-breasted Nuthatch

Sitta canadensis

Size: 4½" (11 cm)

Male: Gray-backed bird with an obvious black eye line and black cap. Rust-red breast and belly.

Female: duller than male and has a gray cap and pale undersides

Juvenile: same as female

Nest: cavity; male and female excavate a cavity or move into a vacant hole; 1 brood per year

Eggs: 5–6; white with red brown markings

Incubation: 11–12 days; female incubates

Fledging: 14–20 days; female and male feed the young

Migration: irruptive; moves around the state during winter in search of food

Food: insects, insect eggs, seeds; comes to seed and suet feeders

Compare: The White-breasted Nuthatch (p. 249) is larger and has a white breast. Yellow-throated Warbler (p. 253) is larger and has a bright-yellow throat.

Stan's Notes: The nuthatch climbs down trunks of trees headfirst, searching for insects. Like a chickadee, it grabs a seed from a feeder and flies off to crack it open. Wedges the seed into a crevice and pounds it open with several sharp blows. The name "Nuthatch" comes from the Middle English moniker *nuthak,* referring to the habit of hacking seeds open. Look for it in mature conifers, where it extracts seeds from pine cones. Excavates a cavity or takes an old woodpecker hole or a natural cavity and builds a nest within. An irruptive migrator, common in some winters and scarce in others. Gives a series of nasal "yank-yank-yank" calls.

Brown-headed Nuthatch

Sitta pusilla

YEAR-ROUND

Size: 4½" (11 cm)

Male: Gray back and dull-white chin, breast and belly. Brown cap bordered by a black line that extends through eyes. Pale gray spot at the nape of neck, hard to see from a distance.

Female: same as male

Juvenile: same as adult

Nest: cavity; female and male construct; 1 brood per year

Eggs: 3–5; white with dark markings

Incubation: 12–14 days; female incubates

Fledging: 18–19 days; female and male feed the young

Migration: non-migrator

Food: insects, seeds; comes to seed feeders

Compare: Carolina Chickadee (p. 247) is similar in size, but it has a black cap. Look for the brown cap of the Brown-headed Nuthatch.

Stan's Notes: A tiny bird of open pine forest in Louisiana and Mississippi. Like other nuthatches, it feeds by creeping up and down twigs and trunks of trees, looking for insects and insect eggs. Works hard to remove seeds from cones on evergreen trees. Has been known to cache pine seeds for later consumption. Visits seed feeders. A cavity nester, it excavates a cavity, takes an abandoned woodpecker home or uses a nest box. Occasionally an unmated male helper attends to a mated female on the nest. Will stay with mate nearly all year, defending a very small territory.

Carolina Chickadee

Poecile carolinensis

Size: 5" (13 cm)

Male: Mostly gray with a black cap and chin. White face and chest with a tan belly. Darker-gray tail.

Female: same as male

Juvenile: same as adult

Nest: cavity; female and male build or excavate; 1–2 broods per year

Eggs: 5–7; white with reddish brown markings

Incubation: 11–12 days; female and male incubate

Fledging: 13–17 days; female and male feed the young

Migration: non-migrator

Food: insects, seeds, fruit; comes to seed and suet feeders

Compare: Brown-headed Nuthatch (p. 245) is similar in size, but it lacks the black cap and chin. Tufted Titmouse (p. 257) is a close relative, but it has an erect crest and lacks the black cap and chin.

Stan's Notes: A common year-round bird in Louisiana and Mississippi. One of the first birds to use a newly placed feeder. Flies to a feeder, grabs a seed and carries it to a branch. To get to the meat inside, it holds the seed down with its feet and hammers the shell open with its bill. Returns for another seed. A friendly bird. Can be tamed and hand-fed. Attracted with a nest box that has a 1¼-inch entrance hole. Female gives a loud snake-like hiss if disturbed on the nest. Often seen with other birds (mixed flock) in winter. Song is a high, fast "chika-dee-dee-dee-dee."

male

female

White-breasted Nuthatch
Sitta carolinensis

YEAR-ROUND

Size: 5–6" (13–15 cm)

Male: Slate gray with a white face, breast and belly. Large white patch on the rump. Black cap and nape. Bill is long and thin, slightly upturned. Chestnut undertail.

Female: similar to male, but has a gray cap and nape

Juvenile: similar to female

Nest: cavity; female and male build a nest within; 1 brood per year

Eggs: 5–7; white with brown markings

Incubation: 11–12 days; female incubates

Fledging: 13–14 days; female and male feed the young

Migration: non-migrator

Food: insects, insect eggs, seeds; comes to seed and suet feeders

Compare: The Red-breasted Nuthatch (p. 243) is smaller and has a rust-red belly and distinctive black eye line. Look for the white breast to help identify the White-breasted Nuthatch.

Stan's Notes: The nuthatch hops headfirst down trees, looking for insects missed by birds climbing up. Its climbing agility is due to an extra-long hind toe claw, or nail, that is nearly twice the size of its front claws. "Nuthatch," from the Middle English *nuthak*, refers to the bird's habit of wedging a seed in a crevice and hacking it open. Often seen in flocks with chickadees and Downy Woodpeckers. Mates stay together year-round, defending a small territory. Gives a characteristic "whi-whi-whi-whi" spring call during February and March. One of 17 worldwide nuthatch species.

male

female

first
winter

Yellow-rumped Warbler
Setophaga coronata

WINTER

Size: 5–6" (13–15 cm)

Male: Slate gray with black streaking on the chest. Yellow patches on the head, flanks and rump. White chin and belly. Two white wing bars.

Female: duller gray than the male, mixed with brown

Juvenile: first winter is similar to the adult female

Nest: cup; female builds; 2 broods per year

Eggs: 4–5; white with brown markings

Incubation: 12–13 days; female incubates

Fledging: 10–12 days; female and male feed young

Migration: complete, to Louisiana, Mississippi, other southern states, Mexico and Central America

Food: insects, berries; visits suet feeders in spring

Compare: Male Common Yellowthroat (p. 361) has a yellow breast and a very distinctive black mask. Prairie Warbler (p. 365) has an olive back with chestnut streaks. The Palm Warbler (p. 367) has a yellow throat and chestnut cap.

Stan's Notes: This is one of our most common winter warblers when the northern breeding birds come to Louisiana and Mississippi for the season. Familiar call is a single robust "chip," heard mostly during migration and winter. Sings a wonderful song in spring. Comes to suet feeders, when insect populations are low. Flits around the upper branches of tall trees. In the fall, the male molts to a dull color similar to the female, but he retains his yellow patches all year. Also called Myrtle Warbler and Audubon's Warbler. Sometimes called Butter-butt due to the yellow patch on its rump. Does not nest in Louisiana and Mississippi.

SUMMER

Yellow-throated Warbler
Setophaga dominica

Size: 5½" (14 cm)

Male: A gray-backed warbler with a bright-yellow throat. Black streaks on white belly. A white spot on neck and white eyebrows.

Female: same as male, but duller with browner back

Juvenile: similar to female

Nest: cup; female and male construct; 1–2 broods per year

Eggs: 4–5; gray with dark markings

Incubation: 12–13 days; female incubates

Fledging: 10–12 days; female and male feed young

Migration: complete, to southern coastal states, Mexico and Central America

Food: insects

Compare: The Yellow-rumped Warbler (p. 251) has yellow on the rump, flanks and head and lacks a yellow throat. The Prairie Warbler (p. 365) is slightly smaller and lacks Yellow-throated's gray back. Palm Warbler (p. 367) has a chestnut cap.

Stan's Notes: A common nesting bird and migrator in Louisiana and Mississippi. Among the first of the warblers nesting in the state, usually around mid-April. Prefers cypress and oak woodlands. Finds food by creeping along and searching underneath vertical surfaces such as tree bark. Highly attracted to water, it is often seen bathing in any puddle of water. In some, the white eyebrows are tinged yellow. It is rarely a Brown-headed Cowbird host.

female
p. 139

male

Dark-eyed Junco
Junco hyemalis

Size: 5½" (14 cm)

Male: Plump, dark-eyed bird with a slate-gray-to-charcoal chest, head and back. White belly. Pink bill. White outer tail feathers appear like a white V in flight.

Female: round with brown plumage

Juvenile: similar to female, with streaking on the breast and head

Nest: cup; female and male build; 2 broods per year

Eggs: 3–5; white with reddish brown markings

Incubation: 12–13 days; female incubates

Fledging: 10–13 days; male and female feed the young

Migration: complete, to Louisiana, Mississippi, other southern states, throughout the U.S.

Food: seeds, insects; visits ground and seed feeders

Compare: Rarely confused with any other bird. Look for the pink bill and small flocks feeding under feeders to identify the male Dark-eyed Junco.

Stan's Notes: A common winter bird in Louisiana and Mississippi. Migrates from Canada to Louisiana and Mississippi. Adheres to a rigid social hierarchy, with dominant birds chasing the less dominant ones. Look for the white outer tail feathers flashing in flight. Often seen in small flocks on the ground, where it uses its feet to simultaneously "double-scratch" to expose seeds and insects. Eats many weed seeds. Nests in a wide variety of wooded habitats. Several subspecies of Dark-eyed Junco were previously considered to be separate species. Males don't go as far south as females in winter.

YEAR-ROUND

Tufted Titmouse
Baeolophus bicolor

Size: 6" (15 cm)

Male: Slate gray with a white chest and belly. Pointed crest. Rust-brown wash on the flanks. Gray legs and dark eyes.

Female: same as male

Juvenile: same as adult

Nest: cavity; female lines an old woodpecker cavity; 2 broods per year

Eggs: 5–7; white with brown markings

Incubation: 13–14 days; female incubates

Fledging: 15–18 days; female and male feed the young

Migration: non-migrator

Food: insects, seeds, fruit; will come to seed and suet feeders

Compare: The Carolina Chickadee (p. 247) is a close relative but is smaller and lacks a crest. The White-breasted Nuthatch (p. 249) has a rust-brown undertail. The Brown-headed Nuthatch (p. 245) is smaller and has a brown cap.

Stan's Notes: A common feeder bird that can be attracted with an offering of black oil sunflower seeds or suet. Can also be attracted with a nest box. Well known for its "peter-peter-peter" call, which it quickly repeats. Notorious for pulling hair from sleeping dogs, cats and squirrels to line its nest. Male feeds female during courtship and nesting. The prefix *tit* in the common name comes from a Scandinavian word meaning "little." Suffix *mouse* is derived from the Old English word *mase*, meaning "bird." Simply translated, it is a "small bird."

breeding
p. 149

winter

Least Sandpiper
Calidris minutilla

Size: 6" (15 cm)

Male: Winter plumage is overall gray to light brown, with a distinct brown breast band and white belly. Light-gray eyebrows and short, thin, down-curved black bill. Dull-yellow legs.

Female: same as male

Juvenile: similar to winter adult, but buff-brown and lacks the breast band

Nest: ground; male and female construct; 1 brood per year

Eggs: 3–4; olive with dark markings

Incubation: 19–23 days; male and female incubate

Fledging: 25–28 days; male and female feed the young

Migration: complete, to Louisiana, Mississippi, other southern coastal states, Mexico, and Central and South America

Food: aquatic and terrestrial insects, seeds

Compare: The smallest of sandpipers. Often confused with winter Western Sandpiper (p. 261). Least Sandpiper's yellow legs differentiate it from other tiny sandpipers. The short, thin, down-curved bill also helps to identify.

Stan's Notes: A winter resident across Louisiana and Mississippi. This is a tiny, tame sandpiper that can be approached without scaring it. It is the smallest of peeps (sandpipers), nesting on the tundra in northern regions of Canada and Alaska. Prefers the grassy flats of saltwater and freshwater ponds. Its yellow legs can be hard to see in water, poor light or when covered with mud. Most other small shorebirds have black legs and feet.

breeding
p. 151

winter

Western Sandpiper
Calidris mauri

Size: 6½" (16 cm)

Male: Winter plumage is dull gray to light brown overall with a white belly and eyebrows. Black legs. Narrow bill that droops near tip.

Female: same as male

Juvenile: similar to breeding adult, bright buff-brown on the back only

Nest: ground; male and female construct; 1 brood per year

Eggs: 2–4; light brown with dark markings

Incubation: 20–22 days; male and female incubate

Fledging: 19–21 days; male and female feed the young

Migration: complete, to Louisiana, Mississippi, other southern coastal states, Mexico, and Central and South America

Food: aquatic and terrestrial insects

Compare: The winter Least Sandpiper (p. 259) is similar but lacks black legs. Look for black legs and a longer bill that droops slightly.

Stan's Notes: A migrant and winter resident across Louisiana and Mississippi. Winters in coastal states from Delaware to California. Nests on the ground in large "loose" colonies on the tundra of northern coastal Alaska. Adults leave their breeding grounds several weeks before the young. Some obtain their breeding plumage before leaving in spring. Feeds on insects at the water's edge, sometimes immersing its head. Young leave the nest (precocial) within a few hours after hatching. Female leaves and the male tends the hatchlings.

Eastern Phoebe
Sayornis phoebe

**YEAR-ROUND
WINTER**

Size: 7" (18 cm)

Male: Plain gray with slightly darker wings, a light-olive belly and a thin, dark bill.

Female: same as male

Juvenile: same as adults

Nest: cup; female builds; 2 broods per year

Eggs: 4–5; white without markings

Incubation: 15–16 days; female incubates

Fledging: 15–16 days; male and female feed the young

Migration: complete to non-migrator in Louisiana and Mississippi, to southern states and Mexico

Food: insects

Compare: The Gray Catbird (p. 275) has a black crown and a chestnut patch under its tail. The Eastern Phoebe lacks any distinctive markings. Listen for its well-enunciated "fee-bee" call and look for the hawking and tail-pumping behaviors to help identify this bird.

Stan's Notes: A sparrow-size bird that often perches on the end of a dead branch. Found in forests, yards and farms. In a process called hawking, it waits for a passing insect. When a bug flies near, it launches out to catch it and then returns to the same branch. It has a distinctive habit of pumping its tail up and down while perching. Builds nest beneath the eaves of houses, under bridges or in other sheltered spots. Uses mud, grass and moss for nest materials and hair (and sometimes feathers) for the lining. The common name is derived from its distinct "fee-bee" call, which it repeats over and over from the top of dead branches.

SUMMER

Great Crested Flycatcher
Myiarchus crinitus

Size: 8" (20 cm)

Male: Gray head with a prominent crest. Gray back and throat. Yellow from the belly to the base of a reddish-brown tail. Lower bill is yellow at the base.

Female: same as male

Juvenile: same as adults

Nest: cavity; female and male construct; 1 brood per year

Eggs: 4–6; white or buff with brown markings

Incubation: 13–15 days; female incubates

Fledging: 14–21 days; female and male feed the young

Migration: complete, to Mexico and Central America

Food: insects, fruit

Compare: The Eastern Kingbird (p. 267) has a white band across its tail. The Eastern Phoebe (p. 263) is similar, but it lacks a crest and yellow belly. Look for the crest to identify the Flycatcher.

Stan's Notes: Breeds throughout both states. Common in almost any wooded area during the summer, it lives high up in trees, rarely coming to the ground. Gleans insects from tree leaves. Often heard before seen. "Great Crested" refers to the set of extra-long feathers on top of its head (crest), which the bird raises when alert or agitated, like the Northern Cardinal. Nests in an old woodpecker hole but can be attracted with a man-made nest box that has an entrance hole 1½–2½ inches (4–6 cm) in diameter. Often stuffs the cavity with a collection of fur, feathers, string and snake skins. Breeds throughout both states.

Eastern Kingbird
Tyrannus tyrannus

SUMMER

Size: 8" (20 cm)

Male: Mostly gray and black with a white chin and belly. Black head and tail with a distinct white band on the tip of the tail. Concealed red crown, rarely seen.

Female: same as male

Juvenile: same as adults

Nest: cup; male and female build; 1 brood per year

Eggs: 3–4; white with brown markings

Incubation: 16–18 days; female incubates

Fledging: 16–18 days; female and male feed the young

Migration: complete, to Mexico, Central America and South America

Food: insects, fruit

Compare: The American Robin (p. 279) is larger and has a rust-red breast. The Eastern Phoebe (p. 263) is smaller and has an olive-green belly. Look for the white tail band to identify the Kingbird.

Stan's Notes: A common summer resident seen in open fields and prairies. Autumn migration begins in late August and early September, with groups of up to 20 birds migrating together. Returns to the mating ground in spring, where pairs defend their territory. Seems to be unafraid of other birds and chases larger birds. Given the common name "King" for its bold attitude and behavior. In a hunting technique known as hawking, it perches on a branch and watches for insects, flies out to catch one, and then returns to the same perch. Swoops from perch to perch when hunting. Becomes very vocal during late summer, when family members call back and forth to one another while hunting for insects.

breeding
p. 169

winter

Sanderling
Calidris alba

MIGRATION
WINTER

Size: 8" (20 cm)

Male: The lightest sandpiper on the beach during winter. Winter plumage has gray head and back and white belly. Black legs and bill. White wing stripe, seen only in flight.

Female: same as male

Juvenile: spotty black on the head and back with a white belly; black legs and bill

Nest: ground; male builds; 1–2 broods per year

Eggs: 3–4; greenish olive with brown markings

Incubation: 24–30 days; male and female incubate

Fledging: 16–17 days; female and male feed the young

Migration: complete, to Louisiana, Mississippi, other southern coastal states, Mexico

Food: insects

Compare: Breeding Spotted Sandpiper (p. 167) is the same size, but it has black spots on the breast and belly. Winter Spotted is similar but has a light-colored bill and yellow legs.

Stan's Notes: A migrant and winter shorebird in Louisiana and Mississippi, but mostly seen in its gray winter plumage from August to April. Seen in groups on sandy beaches, running out with each retreating wave to feed. Look for a flash of white on wings in flight. Sometimes a female mates with several males (polyandry), resulting in males and the female incubating separate nests. Both sexes perform a distraction display if threatened. Rests by standing on one leg and tucking the other into its belly feathers. Often hops away on one leg, moving away from pedestrians on the beach. Surveys show a large decline in numbers since the 1970s.

breeding
p. 171

winter

Dunlin
Calidris alpina

MIGRATION
WINTER

Size: 8–9" (20–23 cm)

Male: Winter adult has a brownish-gray back with a light-gray chest and white belly. Stout bill curves slightly downward at tip. Black legs.

Female: slightly larger than the male, with a longer bill

Juvenile: slightly rusty back with a spotty chest

Nest: ground; male and female construct; 1 brood per year

Eggs: 2–4; olive-buff or blue-green with red-brown markings

Incubation: 21–22 days; male incubates during the day, female incubates at night

Fledging: 19–21 days; male feeds the young, female often leaves before the young fledge

Migration: complete, to Louisiana, Mississippi, other southern coastal states, Mexico, Central America

Food: insects

Compare: Winter Sanderling (p. 269) is a similar size but is lighter and lacks the long down-turned bill.

Stan's Notes: Usually seen in gray winter plumage from August to early May. Breeding plumage is more commonly seen in the spring. Flights include heights of up to 100 feet (30 m) with brief gliding alternating with shallow flutters, and a rhythmic, repeating song. Huge flocks fly synchronously, with birds twisting and turning, flashing light and dark undersides. Males tend to fly farther south in winter than females. Doesn't nest in Louisiana or Mississippi.

red morph

gray morph

YEAR-ROUND

Eastern Screech-Owl
Megascops asio

Size: 8–10" (20–25 cm); up to 2' wingspan

Male: Small "eared" owl that occurs in different colorations. Gray morph is mottled gray and white. Red morph is mottled rust and white. Short wings. Bright-yellow eyes.

Female: same as male but slightly larger

Juvenile: lighter color than adults of the same morph and usually lacks ear tufts

Nest: cavity, old woodpecker cavity or man-made nest box; does not add any nesting material; 1 brood per year

Eggs: 4–5; white without markings

Incubation: 25–26 days; female incubates, male feeds the female during incubation

Fledging: 26–27 days; male and female feed the young

Migration: non-migrator; moves around in winter

Food: large insects, small mammals, birds, snakes

Compare: This is the only small owl in Louisiana and Mississippi with ear tufts. Can be gray or rust in color. The Eastern Screech-Owl is hard to confuse with its much larger cousin, the Great Horned Owl (p. 231).

Stan's Notes: Commonly found in forests that have suitable natural cavities for nesting and roosting. Active from dusk to dawn. Usually gives a tremulous, descending trill, like a sound effect in a scary movie. Seldom gives a screeching call. Often seen sunning itself at a nest-box hole during winter. Mates may have a long-term pair bond and may roost together at night. Excellent hearing and eyesight. The gray morph is more common than the red.

Gray Catbird
Dumetella carolinensis

YEAR-ROUND
SUMMER
WINTER

Size: 9" (22.5 cm)

Male: Handsome slate-gray bird with a black crown and a long, thin, black bill. Often lifts up its tail, exposing a chestnut patch beneath.

Female: same as male

Juvenile: same as adults

Nest: cup; female and male build; 2 broods per year

Eggs: 4–6; blue-green without markings

Incubation: 12–13 days; female incubates

Fledging: 10–11 days; female and male feed young

Migration: complete to non-migrator in Louisiana and Mississippi

Food: insects, occasional fruit; visits suet feeders

Compare: The Eastern Phoebe (p. 263) is smaller and has an olive belly. The Eastern Kingbird (p. 267) is similar in size but has a white belly and a white band across its tail. To identify the Gray Catbird, look for the black crown and chestnut patch under the tail.

Stan's Notes: A secretive bird, more often heard than seen. The Chippewa Indians gave it a name that means "the bird that cries with grief" due to its raspy call. Called "Catbird" because the sound is like the meowing of a house cat. Often mimics other birds, rarely repeating the same phrases. Found in forest edges, backyards and parks. Builds its nest with small twigs. Nests in thick shrubs and quickly flies back into shrubs if approached. If a cowbird lays an egg in its nest, the catbird will quickly break it and eject it.

Loggerhead Shrike
Lanius ludovicianus

YEAR-ROUND

Size: 9" (22.5 cm)

Male: Gray head and back and a white chin, breast and belly. Black wings, tail, legs and feet. Black mask across the eyes and a black bill with a hooked tip. White wing patches, seen in flight.

Female: same as male

Juvenile: dull version of adult

Nest: cup; male and female construct; 1–2 broods per year

Eggs: 4–7; off-white with dark markings

Incubation: 16–17 days; female incubates

Fledging: 17–21 days; female and male feed the young

Migration: non-migrator in Louisiana and Mississippi

Food: insects, lizards, small mammals, frogs

Compare: The Northern Mockingbird (p. 281) has a similar color pattern, but it lacks the black mask. The Cedar Waxwing (p. 161) has a black mask, but it is a brown bird, not gray and black like the Loggerhead Shrike.

Stan's Notes: The Loggerhead is a songbird that acts like a bird of prey. Known for skewering prey on barbed wire fences, thorns and other sharp objects to store or hold still while tearing apart to eat, hence its other common name, Butcher Bird. Feet are too weak to hold the prey it eats. Loggerheads from northern states enter Louisiana and Mississippi in winter, swelling populations. Breeding bird surveys indicate declining populations in the Great Plains due to pesticides killing its major food source—grasshoppers.

male

female

American Robin
Turdus migratorius

YEAR-ROUND

Size: 9–11" (23–28 cm)

Male: Familiar gray bird with a dark rust-red breast and a nearly black head and tail. White chin with black streaks. White eye-ring.

Female: similar to male, with a duller rust-red breast and a gray head

Juvenile: similar to female, with a speckled breast and brown back

Nest: cup; female builds with help from the male; 2–3 broods per year

Eggs: 4–7; pale blue without markings

Incubation: 12–14 days; female incubates

Fledging: 14–16 days; female and male feed the young

Migration: non-migrator in Louisiana and Mississippi

Food: insects, fruit, berries, earthworms

Compare: Familiar bird to all. To differentiate the male from the female, compare the nearly black head and rust-red chest of the male with the gray head and duller chest of the female.

Stan's Notes: Common year-round resident in Louisiana and Mississippi; resident birds are joined by northern migratory robins during winter. Can be heard all night in spring. City robins sing louder than country robins in order to hear one another over traffic and noise. A robin isn't listening for worms when it turns its head to one side. It is focusing its sight out of one eye to look for dirt moving, which is caused by worms moving. Territorial, often fighting its reflection in a window.

displaying

YEAR-ROUND

Northern Mockingbird
Mimus polyglottos

Size: 10" (25 cm)

Male: Silvery-gray head and back with a light-gray breast and belly. White wing patches, seen in flight or during display. Tail mostly black with white outer tail feathers. Black bill.

Female: same as male

Juvenile: dull gray with a heavily streaked breast and a gray bill

Nest: cup; female and male construct; 2 broods per year, sometimes more

Eggs: 3–5; blue-green with brown markings

Incubation: 12–13 days; female incubates

Fledging: 11–13 days; female and male feed the young

Migration: non-migrator in Louisiana and Mississippi

Food: insects, fruit

Compare: Loggerhead Shrike (p. 277) has a similar color pattern, but it is stockier, has a black mask and perches in more-open places. The Gray Catbird (p. 275) is slate gray and lacks wing patches.

Stan's Notes: A very animated bird. Performs an elaborate mating dance. Facing each other with heads and tails erect, pairs will run toward each other, flashing their white wing patches, and then retreat to cover nearby. Thought to flash the wing patches to scare up insects when hunting. Sits for long periods on top of shrubs. Imitates other birds (vocal mimicry); hence the common name. Young males often sing at night. Often unafraid of people, allowing for close observation.

breeding
p. 73

winter

Black-bellied Plover
Pluvialis squatarola

MIGRATION
WINTER

Size: 11–12" (28–30 cm)

Male: Winter plumage is uniform light gray with a white belly and breast. Faint white eyebrow mark. Black legs and bill.

Female: less black on belly and breast than male

Juvenile: grayer than adults, with much less black

Nest: ground; male and female construct; 1 brood per year

Eggs: 3–4; pink or green with black-brown markings

Incubation: 26–27 days; male incubates during the day, female incubates at night

Fledging: 35–45 days; male feeds the young, the young learn quickly to feed themselves

Migration: complete, to Louisiana and Mississippi, the East and Gulf Coasts, West Indies

Food: insects

Compare: The winter Dunlin (p. 271) has a long down-curved bill. Winter Sanderling (p. 269) has a smaller bill. Winter Spotted Sandpiper (p. 167) has a shorter, thicker bill.

Stan's Notes: Males perform a "butterfly" courtship flight to attract females. Female leaves male and young about 12 days after the eggs hatch. Breeds at age 3. A common winter resident along the Louisiana coast, dipping into Mississippi; arrivals start in July and August (fall migration) and leave in April. During flight, in any plumage, displays a white rump and stripe on wings with black axillaries (armpits). Often darts across the ground to grab an insect and run. Can be very common on the beach during winter.

soaring

juvenile

Sharp-shinned Hawk
Accipiter striatus

Size: 10–14" (25–36 cm); up to 2' wingspan

Male: Small woodland hawk with a gray back and head and a rust-red chest. Short wings. Long, squared tail and several dark tail bands, with the widest at the end of the tail. Red eyes.

Female: same as male but larger

Juvenile: same size as adults, with a brown back, heavy streaking on the chest and yellow eyes

Nest: platform; female builds; 1 brood per year

Eggs: 4–5; white with brown markings

Incubation: 32–35 days; female incubates

Fledging: 24–27 days; female and male feed the young

Migration: ccomplete, to southern states, Mexico and Central America; non-migrator in part of Mississippi

Food: birds, small mammals

Compare: Cooper's Hawk (p. 295) is larger and has a larger head, a slightly longer neck and a rounded tail. Red-shouldered Hawk (p. 203) is larger and has a reddish head and belly. Look for the squared tail to help identify,

Stan's Notes: A common hawk of backyards, parks and woodlands. Constructs its nest with sticks, usually high in a tree. Typically seen swooping in on birds visiting feeders and chasing them as they flee. Its short wingspan and long tail help it to maneuver through thick stands of trees in pursuit of prey. Calls a loud, high-pitched "kik-kik-kik-kik." Named "Sharp-shinned" for the sharp projection (keel) on the leading edge of its shin. In most birds, the tarsus bone is rounded, not sharp.

soaring

juvenile

soaring
juvenile

Mississippi Kite
Ictinia mississippiensis

SUMMER MIGRATION

Size: 12–15" (30–38 cm); up to 2¾' wingspan

Male: Overall gray bird with a paler, nearly white head. Nearly black tail. Dark eye patch surrounding red eyes. Short, hooked gray bill. Yellow legs and feet.

Female: same as male

Juvenile: similar to adult but has a brown chest with vertical white streaks

Nest: platform; female and male build; 1 brood per year

Eggs: 1–2; white without markings

Incubation: 29–32 days; female and male incubate

Fledging: 32–34 days; female and male feed young

Migration: complete, to South America

Food: insects, lizards, small snakes

Compare: Smaller than many other birds of prey. The overall gray appearance with a lighter head makes it easy to identify.

Stan's Notes: A bird of prey that eats mostly large insects. Groups follow livestock, feeding on insects they kick up. Hunts insects by soaring or hovering, catching in flight or diving down. It requires open areas with scattered trees for nesting. Nests in semi-colonies. Mated pairs aggressively defend nest sites. Individuals often stray out of traditional ranges, appearing in northern states and up the East Coast.

YEAR-ROUND

Eurasian Collared-Dove
Streptopelia decaocto

Size: 12½" (32 cm)

Male: Head, neck, breast and belly are gray to tan. Back, wings and tail are slightly darker. Thin black collar with a white border on the nape of the neck. Tail is long and squared.

Female: same as male

Juvenile: similar to adults

Nest: platform; female and male build; 2–3 broods per year

Eggs: 3–5; creamy white without markings

Incubation: 12–14 days; female and male incubate

Fledging: 12–14 days; female and male feed the young

Migration: non-migrator

Food: seeds; will visit ground and seed feeders

Compare: The Mourning Dove (p. 189) is slightly smaller and darker. The Rock Pigeon (p. 291) has colorful iridescent patches. Look for the black collar on the nape and the squared tail to help identify the Eurasian Collared-Dove.

Stan's Notes: A non-native bird. Moved into Florida in the 1980s after inadvertent introduction to the Bahamas; then spread to Mississippi and Louisiana. Reached the northern states in the late 1990s. It has been expanding its range across North America and is predicted to spread just like it did through Europe from Asia. Unknown how this "new" bird will affect populations of the native Mourning Dove. Nearly identical to the Ringed Turtle-Dove, a common pet bird. The dark mark on the back of the neck gave rise to the common name. Look for flashes of white in the tail and dark wing tips when it lands or takes off.

YEAR-ROUND

Rock Pigeon
Columba livia

Size: 13" (33 cm)

Male: No set color pattern. Shades of gray to white with patches of gleaming, iridescent green and blue. Often has a light rump patch.

Female: same as male

Juvenile: same as adults

Nest: platform; female builds; 3–4 broods per year

Eggs: 1–2; white without markings

Incubation: 18–20 days; female and male incubate

Fledging: 25–26 days; female and male feed the young

Migration: non-migrator

Food: seeds

Compare: The Eurasian Collared-Dove (p. 289) has a black collar on the nape. The Mourning Dove (p. 189) is smaller and light brown and lacks the variety of color combinations of the Rock Pigeon.

Stan's Notes: Also known as the Domestic Pigeon. Formerly known as the Rock Dove. Introduced to North America from Europe by the early settlers. Most common around cities and barnyards, where it scratches for seeds. One of the few birds with a wide variety of colors, produced by years of selective breeding while in captivity. Parents feed the young a regurgitated liquid known as crop-milk for the first few days of life. One of the few birds that can drink without tilting its head back. Nests under bridges or on buildings, balconies, barns and sheds. Was once thought to be a nuisance in cities and was poisoned. Now, many cities have Peregrine Falcons (p. 297) feeding on Rock Pigeons, which keeps their numbers in check.

breeding
p. 195

displaying

winter

Willet

Catoptrophorus semipalmatus

Size: 14–16" (36–40 cm)

Male: Winter plumage is gray with a white belly. A distinctive black-and-white wing lining pattern, seen in flight or during display. Gray bill and legs.

Female: same as male

Juvenile: similar to breeding adult, more tan in color

Nest: ground; female builds; 1 brood per year

Eggs: 3–5; olive-green with dark markings

Incubation: 24–28 days; male and female incubate

Fledging: 1–2 days; female and male feed young

Migration: non-migrator in southern parts of Louisiana and Mississippi

Food: insects, small fish, crabs, worms, clams

Compare: Greater Yellowlegs (p. 193) is slightly smaller and has a longer neck and yellow legs. Killdeer (p. 183) has 2 black bands on the neck.

Stan's Notes: A year-round resident in Louisiana and Mississippi. One of the few shorebirds that is seen away from water, sometimes standing on fence posts. It appears a rich, warm brown during the breeding season and rather plain gray during the winter, but it always has a striking black-and-white wing pattern when seen in flight. Uses its black-and-white wing patches to display to its mate. Named after the "pill-will-willet" call it gives during the breeding season. Gives a "kip-kip-kip" alarm call when it takes flight. Nests along the Gulf and East Coasts, in some western states and Canada.

soaring

juvenile

Cooper's Hawk
Accipiter cooperii

YEAR-ROUND WINTER

Size: 14–20" (36–51 cm); up to 3' wingspan

Male: Medium-size hawk with short wings and a long, rounded tail with several black bands. Slate-gray back, rusty breast, dark wing tips. Gray bill with a bright-yellow spot at the base. Dark-red eyes.

Female: similar to male but larger

Juvenile: brown back, brown streaking on the breast, bright-yellow eyes

Nest: platform; male and female construct; 1 brood per year

Eggs: 2–4; greenish with brown markings

Incubation: 32–36 days; female and male incubate

Fledging: 28–32 days; male and female feed the young

Migration: non-migrator to partial in Louisiana and Mississippi; moves around to find food

Food: small birds, mammals

Compare: The Sharp-shinned Hawk (p. 285) is smaller and lighter gray and has a squared tail. Look for the banded, rounded tail to help identify Cooper's Hawk.

Stan's Notes: Found in many habitats, from woodlands to parks and backyards. Stubby wings help it to navigate around trees while it chases small birds. Will ambush prey, flying into heavy brush or even running on the ground in pursuit. Comes to feeders, hunting for birds. Flies with long glides followed by a few quick flaps. The young have gray eyes that turn bright yellow at 1 year and turn dark red later, after 3–5 years. Populations swell in winter, when resident hawks are joined by northern migrants.

juvenile

in-flight
juvenile

in flight

WINTER

Peregrine Falcon
Falco peregrinus

Size: 16–20" (41–51 cm); up to 3¾' wingspan

Male: Dark-gray back and tan-to-white chest. Horizontal bars on belly, legs and undertail. Dark "hood" head marking and wide black mustache. Yellow base of bill and eye-ring. Yellow legs.

Female: similar to male but noticeably larger

Juvenile: overall darker than adults, with heavy streaking on the chest and belly

Nest: ground (scrape) on a cliff edge, tall building, bridge or smokestack; 1 brood per year

Eggs: 3–4; white, some with brown markings

Incubation: 29–32 days; female and male incubate

Fledging: 35–42 days; male and female feed the young

Migration: winters in southern Louisiana and Mississippi

Food: birds (Rock Pigeons in cities, shorebirds and waterfowl in rural areas)

Compare: The American Kestrel (p. 179) is smaller and has 2 vertical black stripes on its face. Look for the dark "hood" head marking and mustache marks to identify the Peregrine Falcon.

Stan's Notes: A wide-bodied raptor that hunts many bird species. The larger females hunt larger prey. Lives in many cities, diving (stooping) on pigeons at speeds of up to 200 miles (322 km) per hour, which knocks them to the ground. Soars with its wings flat, often riding thermals. During courtship, the male brings food to the female and performs aerial displays. Likes to nest on a high ledge or platform with a good view. A solitary nester and monogamous.

female
p. 225

male

soaring

Northern Harrier
Circus hudsonius

WINTER

Size: 18–22" (45–56 cm); up to 4' wingspan

Male: Slender, low-flying hawk. Silver-gray with a large white rump patch and white belly. Long tail with faint narrow bands. Black wing tips. Yellow eyes.

Female: dark-brown back, brown streaking on breast and belly, large white rump patch, thin black tail bands, black wing tips, yellow eyes

Juvenile: similar to female, with an orange breast

Nest: ground; female and male construct; 1 brood per year

Eggs: 4–8; bluish white without markings

Incubation: 31–32 days; female incubates

Fledging: 30–35 days; male and female feed the young

Migration: complete, to Louisiana, Mississippi, other southern states, Mexico and Central America

Food: mice, snakes, insects, small birds

Compare: Slimmer than Red-tailed Hawk (p. 227). Cooper's Hawk (p. 295) has a rusty breast. Look for a low-gliding hawk with a large white rump patch to identify the male Harrier.

Stan's Notes: One of the easiest of hawks to identify. Glides just above the ground, following the contours of the land while searching for prey. Holds its wings just above horizontal, tilting back and forth in the wind, similar to Turkey Vultures. Formerly called the Marsh Hawk due to its habit of hunting over marshes. Feeds and nests on the ground. Will also preen and rest on the ground. Unlike other hawks, mainly uses its hearing to find prey, followed by sight. At any age, has a distinctive owl-like face disk.

juvenile

in flight

Yellow-crowned Night-Heron
Nyctanassa violacea

Size: 24" (60 cm); up to 3½' wingspan

Male: Stocky gray heron with a black head, white cheek patch and yellow-to-white crown. Thick dark bill. Slender yellow legs. Long, thin white plumes extend from the back of head during breeding season.

Female: same as male

Juvenile: brown with white streaks and a dark bill, green legs

Nest: platform; female and male build; 1 brood per year

Eggs: 4–6; light blue without markings

Incubation: 21–25 days; female and male incubate

Fledging: 21–25 days; female and male feed the young

Migration: partial to non-migrator in Louisiana and Mississippi

Food: aquatic insects, fish, crustaceans

Compare: Great Blue Heron (p. 305) is twice as large, has longer legs and lacks a black chin. The distinctively patterned head makes this heron easy to identify.

Stan's Notes: This heron hunts in the evening and early morning, as the common name implies, but it can also be active during the day. Found from coastal mangroves to interior swamps, often hunting fiddler crabs and crayfish. Not uncommon for it to nest in large heron rookeries. Sometimes will nest by itself or in small colonies. Usually seen alone or in small groups. During breeding season, the crown acquires a yellow hue.

in flight

Canada Goose
Branta canadensis

YEAR-ROUND
WINTER

Size: 25–43" (64–109 cm); up to 5½' wingspan

Male: Large gray goose with a black neck and head. White chin and cheek strap.

Female: same as male

Juvenile: same as adults

Nest: platform, on the ground; female builds; 1 brood per year

Eggs: 5–10; white without markings

Incubation: 25–30 days; female incubates

Fledging: 42–55 days; male and female teach the young to feed

Migration: complete, to Louisiana, Mississippi and other southern states; moves around in winter to find food

Food: aquatic plants, insects, seeds

Compare: rarely confused with any other bird

Stan's Notes: Formerly killed off (extirpated) in many areas, it was reintroduced and is now a common resident. Adapting to our changed environment very well. Calls a classic "honk-honk-honk," especially in flight. Flocks fly in a large V when traveling long distances. Winter resident in Louisiana and Mississippi. Begins breeding in the third year. Adults mate for many years. If threatened, they will hiss as a warning. Males stand as sentinels at the edge of their group and will bob their heads and become aggressive if approached. Adults molt their primary flight feathers while raising their young, rendering family groups temporarily flightless. Several subspecies vary in the U.S. Generally eastern groups are paler than western. Their size also varies, decreasing northward. The smallest subspecies is in the Arctic.

in flight

Great Blue Heron
Ardea herodias

Size: 42–48" (107–122 cm); up to 6' wingspan

Male: Tall and gray. Black eyebrows end in long plumes at the back of the head. Long yellow bill. Long feathers at the base of the neck drop down in a kind of necklace. Long legs.

Female: same as male

Juvenile: same as adults, but more brown than gray, with a black crown; lacks plumes

Nest: platform in a colony; male and female build; 1 brood per year

Eggs: 3–5; blue-green without markings

Incubation: 27–28 days; female and male incubate

Fledging: 56–60 days; male and female feed the young

Migration: non-migrator in Louisiana and Mississippi

Food: small fish, frogs, insects, snakes, baby birds

Compare: Tricolored Heron (p. 121) is half the size of the Great Blue Heron and has a white belly. Green Heron (p. 313) is much smaller and has a short neck. Look for the long, yellow bill to help identify the Great Blue Heron.

Stan's Notes: One of the most common herons in Louisiana and Mississippi. Found in open water, from small ponds to large lakes. Stalks small fish in shallow water. Will strike at mice, squirrels and nearly anything it comes across. Red-winged Blackbirds will attack it to stop it from taking their babies out of the nest. In flight, it holds its neck in an S shape and slightly cups its wings, while the legs trail straight out behind. Nests in a colony of up to 100 birds. Nests in trees near or hanging over water. Barks like a dog when startled.

in flight

rusty
stain

in-flight
rusty stain

YEAR-ROUND
MIGRATION
WINTER

Sandhill Crane
Grus canadensis

Size: 42–48" (107–122 cm); up to 7' wingspan

Male: Elegant gray crane with long legs and neck. Wings and body often rust brown from mud staining. Scarlet-red cap. Yellow to red eyes.

Female: same as male

Juvenile: dull brown with yellow eyes; lacks a red cap

Nest: ground; female and male construct; 1 brood per year

Eggs: 2; olive with brown markings

Incubation: 28–32 days; female and male incubate

Fledging: 65 days; female and male feed the young

Migration: complete, to southern coastal states, Mexico; small wintering population in some areas

Food: insects, fruit, worms, plants, amphibians

Compare: Great Blue Heron (p. 305) has a longer bill and holds its neck in an S shape during flight. Look for the scarlet-red cap to help identify the Sandhill Crane.

Stan's Notes: Preens mud into its feathers, staining its plumage rust brown (see insets). Gives a very loud and distinctive rattling call, often heard before the bird is seen. Flight is characteristic, with a faster upstroke, making the wings look like they're flicking in flight. Can fly at heights of over 10,000 feet (3,050 m). Nests on the ground in a large mound of aquatic vegetation. Performs a spectacular mating dance: The birds will face each other, then bow and jump into the air while making loud cackling sounds and flapping their wings. They will also flip sticks and grass into the air during their dance.

male

female

Ruby-throated Hummingbird
Archilochus colubris

SUMMER

Size: 3–3½" (7.5–9 cm)

Male: Tiny iridescent green bird with black throat patch that reflects bright ruby red in sun.

Female: same as male, but lacking the throat patch

Juvenile: same as female

Nest: cup; female builds; 1–2 broods per year

Eggs: 2; white without markings

Incubation: 12–14 days; female incubates

Fledging: 14–18 days; female feeds the young

Migration: complete, to Mexico and Central America

Food: nectar, insects; will come to nectar feeders

Compare: No other bird is as tiny. The Sphinx Moth also hovers at flowers but has clear wings, doesn't hum in flight, moves much slower than the Ruby-throated and can be approached.

Stan's Notes: This is the smallest bird in Louisiana and Mississippi. Can fly straight up, straight down or backward and hover in mid-air. Does not sing but chatters or buzzes to communicate. Weighs about the same as a U.S. penny; it takes about five average-size hummingbirds to equal the weight of one chickadee. The wings create the humming sound. Flaps 50–60 times or more per second when flying at top speed. Breathes 250 times per minute. Heart beats up to 1,260 times per minute. Builds a stretchy nest with plant material and spiderwebs, gluing pieces of lichen to the exterior for camouflage. Attracted to colorful, tubular flowers. Feeds on a combination of nectar and insects. Will extract and eat insects trapped in spiderwebs. A long-distance migrator, wintering in the tropics of Mexico and Central America.

male

female

Painted Bunting
Passerina ciris

SUMMER

Size: 5½" (14 cm)

Male: An amazing combination of colors. A green back, deep-blue head and orange chest and belly. Dark wings and tail.

Female: bright green above, light green below

Juvenile: drab version of the female with only some small spots of green

Nest: cup; female and male construct; 1–2 broods per year

Eggs: 3–5; white without markings

Incubation: 11–12 days; female incubates

Fledging: 12–14 days; female and male feed the young

Migration: complete, to Florida, Bahamas, Cuba, Mexico and Central America

Food: seeds, insects; will visit seed feeders

Compare: No other bird can compare to the male's striking colors. The female is uniquely green and rarely confused with any other bird.

Stan's Notes: A wonderful bunting of backyard gardens, woodland edges and along brushy roads. Visits seed feeders in wooded yards. Well known for its loud, clear and varied warbling phrases. Cup nest, made of grass and lined with animal hair, is usually in a deep, tangled mass of vines. A common cowbird host, this unfortunately often results in raising the cowbird young and not its own. Nests across Louisiana and Mississippi. Often captured in Central America and sold as a caged bird; both activities are illegal in the U.S. and should not be supported.

in flight

YEAR-ROUND
SUMMER

Green Heron
Butorides virescens

Size: 16–22" (41–56 cm)

Male: Short and stocky. Blue-green back and rust-red neck and breast. Dark-green crest. Short legs are normally yellow but turn bright orange during the breeding season.

Female: same as male

Juvenile: similar to adults, with a bluish-gray back and white-streaked breast and neck

Nest: platform; female and male build; 2 broods

Eggs: 2–4; light green without markings

Incubation: 21–25 days; female and male incubate

Fledging: 35–36 days; female and male feed the young

Migration: complete, to southern states, Mexico and Central and South America; non-migrator in southern Louisiana

Food: small fish, aquatic insects, small amphibians, aquatic plants

Compare: Tricolored Heron (p. 121) and Great Blue Heron (p. 305) are larger and have a long neck. Green Heron lacks the long neck of most other herons. Look for a small heron with a dark green back and crest stalking wetlands.

Stan's Notes: Often gives an explosive, rasping "skyew" call when startled. Waits on the shore or wades stealthily, hunting for small fish, aquatic insects and small amphibians. Places an object, such as an insect, on the water's surface to attract fish to catch. Nests in a tall tree, often a short distance from the water. The nest can be very high up in the tree. Babies give a loud ticking sound, like the ticktock of a clock.

female
p. 215

male

Wood Duck
Aix sponsa

YEAR-ROUND

Size: 17–20" (43–51 cm)

Male: Small, highly ornamented dabbling duck. Mostly green head and crest patterned with black and white. Rusty chest and a white belly. Red eyes.

Female: brown duck with a bright-white eye-ring, not-so-obvious crest and blue patch on wings (speculum), often hidden

Juvenile: similar to female

Nest: cavity; female lines an old woodpecker cavity or a nest box in a tree; 1 brood per year

Eggs: 10–15; creamy white without markings

Incubation: 28–36 days; female incubates

Fledging: 56–68 days; female teaches the young to feed

Migration: non-migrator in Louisiana and Mississippi

Food: aquatic insects, plants, seeds

Compare: Male Northern Shoveler (p. 319) is larger and has a long, wide bill.

Stan's Notes: A common duck of quiet, shallow backwater ponds. Nearly went extinct around 1900 due to overhunting, but it's doing well now. Nests in a tree cavity or a nest box in a tree. Seen flying in forests or perching on high branches. Female takes off with a loud squealing call and enters the nest cavity from full flight. Lays some eggs in a neighboring nest (egg dumping), resulting in more than 20 eggs in some clutches. Hatchlings stay in the nest for 24 hours, then jump from as high as 60 feet (18 m) to the ground or water to follow their mother. They never return to the nest. Populations increase during winter with northern migrants.

female
p. 219

male

WINTER

Mallard
Anas platyrhynchos

Size: 19–21" (48–53 cm)

Male: Large, bulbous green head, white necklace and rust-brown or chestnut chest. Gray-and-white sides. Yellow bill. Orange legs and feet.

Female: brown with an orange-and-black bill and blue-and-white wing mark (speculum)

Juvenile: same as female but with a yellow bill

Nest: ground; female builds; 1 brood per year

Eggs: 7–10; greenish to whitish, unmarked

Incubation: 26–30 days; female incubates

Fledging: 42–52 days; female leads the young to food

Migration: complete, to Louisiana and Mississippi, other southern states

Food: seeds, plants, aquatic insects; will come to ground feeders offering corn

Compare: Most people recognize this common duck. Male Northern Shoveler (p. 319) has a white chest with rust on sides and a dark spoon-shaped bill.

Stan's Notes: A familiar dabbling duck of lakes and ponds. Also found in rivers, streams and some backyards. Tips forward to feed on vegetation on the bottom of shallow water. The name "Mallard" comes from the Latin word *masculus,* meaning "male," referring to the male's habit of taking no part in raising the young. Male and female have white underwings and white tails, but only the male has black central tail feathers that curl upward. Unlike the female, the male doesn't quack. Returns to its birthplace each year.

female
p. 221

male

WINTER

Northern Shoveler
Anas clypeata

Size: 19–21" (48–53 cm)

Male: Medium-size duck with an iridescent green head, rust sides, white chest. Extraordinarily large, spoon-shaped bill, almost always held pointed toward the water.

Female: brown and black, with a green wing patch (speculum) and large, spoon-shaped bill

Juvenile: same as female

Nest: ground; female builds; 1 brood per year

Eggs: 9–12; olive without markings

Incubation: 22–25 days; female incubates

Fledging: 30–60 days; female leads the young to food

Migration: complete, to Louisiana, Mississippi, other southern states, Mexico and Central America

Food: aquatic insects, plants

Compare: The male Mallard (p. 317) is similar but lacks the large, spoon-shaped bill. The male Wood Duck (p. 315) is smaller and has a crest.

Stan's Notes: One of several species of shovelers. Called "Shoveler" due to the peculiar, shovel-like shape of its bill. Given the common name "Northern" because it is the only species of these ducks in North America. Seen in shallow wetlands, ponds and small lakes in flocks of 5–10 birds. Flocks fly in tight formation. Swims low in water, pointing its large bill toward the water as if it's too heavy to lift. Usually swims in tight circles while feeding. Feeds mainly by filtering tiny aquatic insects and plants from the surface of the water with its bill. Female gathers plant material and forms it into a nest a short distance from the water. A winter resident in Louisiana and Mississippi that arrives in September and leaves in April.

female
p. 363

male

SUMMER
MIGRATION

American Redstart
Setophaga ruticilla

Size: 5" (13 cm)

Male: Striking black warbler with orange patches on the sides, wings and tail. White belly.

Female: olive-brown with yellow patches on the sides, wings and tail, white belly

Juvenile: same as female; male attains orange tinges in the second year

Nest: cup; female builds; 1 brood per year

Eggs: 3–5; off-white with brown markings

Incubation: 12 days; female incubates

Fledging: 9 days; female and male feed the young

Migration: complete, to Mexico, Central America and South America

Food: insects, seeds, occasionally berries

Compare: The male Baltimore Oriole (p. 323) and male Red-winged Blackbird (p. 31) are much larger. The male American Redstart is the only small black-and-orange bird flitting around the top of trees.

Stan's Notes: A common widespread warbler in Louisiana and Mississippi during summer and migration. Prefers large, unbroken tracts of forest. Found in woodlands, parks and yards and at forest edges. Appears hyperactive when it feeds, hovering and darting back and forth to glean insects from leaves. Often droops wings and fans tail before launching out to catch an insect. Look for the flashing black-and-orange colors of the male high up in trees. First-year males have yellow markings and look like the females. Sings a high-pitched song that builds in intensity and then suddenly ends.

female
p. 371

male

Baltimore Oriole

Icterus galbula

Size: 7–8" (18–20 cm)

Male: Flaming orange with a black head and back. White-and-orange wing bars. Orange-and-black tail. Gray bill and dark eyes.

Female: pale yellow with orange tones, gray-brown wings, white wing bars, gray bill, dark eyes

Juvenile: same as female

Nest: pendulous; female builds; 1 brood per year

Eggs: 4–5; bluish with brown markings

Incubation: 12–14 days; female incubates

Fledging: 12–14 days; female and male feed the young

Migration: complete, to Mexico, Central America and South America

Food: insects, fruit, nectar; comes to nectar, orange-half, and grape-jelly feeders

Compare: The male Orchard Oriole (p. 325) is much darker orange. Look for the flaming orange to identify the male Baltimore Oriole. Male American Redstart (p. 321) is smaller and has more black than orange.

Stan's Notes: A fantastic songster, often heard before seen. Easily attracted to a feeder that offers sugar water (nectar), orange halves, or grape jelly. Parents bring their young to feeders. Sits at the tops of trees, feeding on caterpillars. Female builds a sock-like nest at the outermost branches of tall trees. Prefers parks, yards and forests and often returns to the same area year after year. Young males turn orange-and-black at 1½ years of age. Some of the last birds to arrive in spring and first to leave in fall. Seen during migration and summer.

female
p. 373

male

first-year
male

SUMMER

Orchard Oriole
Icterus spurius

Size: 7–8" (18–20 cm)

Male: Dark orange with black head, throat, upper back, wings and tail. White wing bar. Bill is long and thin. Gray mark on lower bill.

Female: olive-green back, dull-yellow belly and gray wings with 2 indistinct white wing bars

Juvenile: same as female; first-year male looks like the female, with a black bib

Nest: pendulous; female builds; 1 brood per year

Eggs: 3–5; pale blue to white, brown markings

Incubation: 11–12 days; female and male incubate

Fledging: 11–14 days; female and male feed the young

Migration: complete, to Mexico, Central America and northern South America

Food: insects, fruit, nectar; comes to nectar, orange-half and grape-jelly feeders

Compare: Male Northern Cardinal (p. 333) is larger and red. Male Summer Tanager (p. 331) lacks the black head of the male Orchard Oriole. Male Baltimore Oriole (p. 323) is a brighter orange.

Stan's Notes: Named "Orchard" for its preference for orchards. Also likes open woods. Eats insects until wild fruit starts to ripen. Often nests alone; sometimes nests in small colonies. Parents bring their young to bird feeding stations after they fledge. Many people don't see these birds at their feeders very much during the summer and think they have left, but the birds are still there, hunting for insects to feed to their young. Often migrates with Baltimore Orioles.

male

female
p. 131

yellow
male

House Finch

Haemorhous mexicanus

YEAR-ROUND

Size: 5" (13 cm)

Male: Small finch with a red-to-orange face, throat, chest and rump. Brown cap. Brown marking behind eyes. White belly with brown streaks. Brown wings with white streaks.

Female: brown with a heavily streaked white chest

Juvenile: similar to female

Nest: cup, sometimes in cavities; female builds; 2 broods per year

Eggs: 4–5; pale blue, lightly marked

Incubation: 12–14 days; female incubates

Fledging: 15–19 days; female and male feed the young

Migration: non-migrator; will move around to find food

Food: seeds, fruit, leaf buds; visits seed feeders and feeders that offer grape jelly

Compare: The male Purple Finch (p. 329) has a red cap. Look for the brown cap and streaked belly to help identify the male House Finch.

Stan's Notes: Can be a common bird at your feeders. Very social, visiting feeders in small flocks. Likes to nest in hanging flower baskets. Male sings a loud, cheerful warbling song. It was originally introduced to Long Island, New York, from the western U.S. in the 1940s and is now found throughout the country. Suffers from a disease that causes the eyes to crust, resulting in blindness and death. Rarely, males are yellow (inset), perhaps due to poor diet.

female
p. 145

male

WINTER

Purple Finch
Haemorhous purpureus

Size: 6" (15 cm)

Male: Raspberry-red head, cap, chest, back and rump. Brownish wings and tail. Large bill.

Female: heavily streaked brown-and-white bird with bold white eyebrows

Juvenile: same as female

Nest: cup; female and male build; 1 brood per year

Eggs: 4–5; greenish blue with brown markings

Incubation: 12–13 days; female incubates

Fledging: 13–14 days; female and male feed the young

Migration: irruptive; moves around in search of food

Food: seeds, insects, fruit; comes to seed feeders

Compare: The male House Finch (p. 327) has a brown cap and a streaked belly. Look for the raspberry cap to help identify the male Purple Finch.

Stan's Notes: Seen only during the winter, when flocks of Purple Finches leave their homes farther north and move around searching for food. Seen in some winters, absent in others. Travels in flocks of up to 50 birds. Visits seed feeders along with House Finches, which makes it hard to tell them apart. Feeds mainly on seeds; ash tree seeds are an important source of food. Found in coniferous forests, mixed woods, woodland edges and suburban backyards. Flies in the typical undulating, up-and-down pattern of finches. Sings a rich, loud song. Gives a distinctive "tic" note only in flight. Male is not purple. The Latin species name *purpureus* means "purple" (or other reddish colors).

female
p. 375

male

Summer Tanager
Piranga rubra

SUMMER

Size: 8" (20 cm)

Male: Bright rosy-red bird with darker red wings.

Female: overall yellow with slightly darker wings

Juvenile: male has patches of red and green over the entire body, female is same as adult female

Nest: cup; female builds; 1–2 broods per year

Eggs: 3–5; pale blue with dark markings

Incubation: 10–12 days; female incubates

Fledging: 12–15 days; female and male feed young

Migration: complete, to Mexico and Central and South America

Food: insects, fruit

Compare: Similar size as the male Northern Cardinal (p. 333), but the male Cardinal has a black mask, large crest and red bill.

Stan's Notes: A distinctive bird of woodlands throughout Louisiana and Mississippi, especially in mixed pine and oak forests. Due to clearing of land for agriculture, populations have decreased for over a century and especially most recently. Returning to the state in late March and with young hatching in late May, some pairs have two broods per year. Most leave the state by November with very few staying for winter. While fruit makes up some of the diet, most of it consists of insects such as bees and wasps. Summer Tanagers unfortunately seem to be parasitized by Brown-headed Cowbirds more than any other nesting bird in Louisiana and Mississippi.

female
p. 173

male

juvenile

Northern Cardinal
Cardinalis cardinalis

YEAR-ROUND

Size: 8–9" (20–23 cm)

Male: Red with a black mask that extends from the face to the throat. Large crest and a large red bill.

Female: buff-brown with a black mask, large reddish bill, and red tinges on the crest and wings

Juvenile: same as female but with a blackish-gray bill

Nest: cup; female builds; 2–3 broods per year

Eggs: 3–4; bluish white with brown markings

Incubation: 12–13 days; female and male incubate

Fledging: 9–10 days; female and male feed the young

Migration: non-migrator

Food: seeds, insects, fruit; comes to seed feeders

Compare: Similar size as the male Summer Tanager (p. 331), but the male Tanager is rosy red. Look for the black mask, large crest and red bill to identify the male Northern Cardinal.

Stan's Notes: A familiar backyard bird. Seen in a variety of habitats, including parks. Usually likes thick vegetation. One of the few species in which both males and females sing. Can be heard all year. Listen for its "whata-cheer-cheer-cheer" territorial call in spring. Watch for a male feeding a female during courtship. The male also feeds the young of the first brood while the female builds a second nest. Territorial in spring, fighting its own reflection in a window or other reflective surface. Non-territorial in winter, gathering in small flocks of up to 20 birds. Makes short flights from cover to cover, often landing on the ground. *Cardinalis* denotes importance, as represented by the red priestly garments of Catholic cardinals.

female
p. 223

male

Canvasback
Aythya valisineria

Size: 20–21" (51–53 cm)

Male: Deep-red head with a sloping forehead that transitions into a long black bill. Red neck. Gray-and-white sides and back. Black chest and tail.

Female: similar to male, but has a brown head, neck and chest, light gray-to-brown sides and a long dark bill

Juvenile: similar to female

Nest: ground; female builds; 1 brood per year

Eggs: 7–9; pale white to gray without markings

Incubation: 24–29 days; female incubates

Fledging: 56–67 days; female leads young to food

Migration: complete, to Louisiana, Mississippi, other southern states and Mexico

Food: aquatic insects, small clams

Compare: The male Lesser Scaup (p. 79) is smaller, lacks the red head and neck of the male Canvasback and has a shorter, light-blue bill.

Stan's Notes: A large inland duck of freshwater lakes, rivers and ponds. Populations declined dramatically in the 1960–80s due to marsh drainage for agriculture. Females return to their birthplace (philopatric) while males disperse to new areas. Will mate during migration or on the breeding grounds. A courting male gives a soft cooing call when displaying and during aerial chases. Male leaves the female after incubation starts. Female takes a new mate every year. Female feeds very little during incubation and will lose up to 70 percent of fat reserves during that time.

in flight

juvenile

YEAR-ROUND
MIGRATION

Roseate Spoonbill
Platalea ajaja

Size: 30–34" (76–86 cm); up to 4' wingspan

Male: An overall pink bird with red highlights. A white neck with a black patch on the back of the head. Heavy, spoon-shaped flat bill. Long red legs.

Female: same as male

Juvenile: pale version of adult

Nest: platform; female and male build; 1 brood per year

Eggs: 1–4; olive green with dark markings

Incubation: 22–23 days; male and female incubate

Fledging: 35–42 days; female and male feed young

Migration: non-migrator, some will move to Florida and Mexico for winter

Food: fish, aquatic insects, snails, worms, leeches

Compare: This is an unmistakable bird. The Roseate Spoonbill is larger than White Ibis (p. 353) and has a long, down-curved orange-to-red bill, unlike the heavy flat bill of the Spoonbill.

Stan's Notes: A year-round Gulf Coast resident. This bird is making a comeback from devastating hunting pressures in the 1800s for its wing feathers, which were used in women's hats and fans. Now habitat destruction is limiting its numbers. Swings its spoon-shaped bill to sift fish and insects from shallow waters. Usually seen in small flocks. Nests in mixed colonies with herons. Related to the ibises.

in flight

Forster's Tern
Sterna forsteri

YEAR-ROUND
MIGRATION
WINTER

Size: 14–15" (36–38 cm); up to 2½' wingspan

Male: White-and-gray tern with a jet-black crown. Leading edge of wings is gray, and trailing edge is white. Characteristic forked tail is long and white. Orange bill with a black tip. Winter plumage lacks the black crown, and the bill becomes nearly entirely black.

Female: same as male

Juvenile: similar to adults but lacks the black crown

Nest: floating platform; female and male build; 1 brood per year

Eggs: 3–5; tan to white with brown markings

Incubation: 23–24 days; female and male incubate

Fledging: 24–26 days; male and female feed the young

Migration: complete, to Louisiana, Mississippi, other southern coastal states, Mexico and Central America; non-migrator in parts of both states

Food: small fish, aquatic insects

Compare: Royal Tern (p. 345) is larger and has a larger orange-red bill. Look for a jet-black crown, orange bill with black tip and white tips of wings to help identify Forster's Tern.

Stan's Notes: Usually seen in small colonies. Catches small fish by diving into water headfirst. Will catch insects in flight. Constructs a platform nest on floating vegetation. Nests in small colonies in shallow-water marshes. Was named after Johann Reinhold Forster, a German naturalist who traveled with Captain Cook in 1772. Populations along the coast increase with northern migrants during winter.

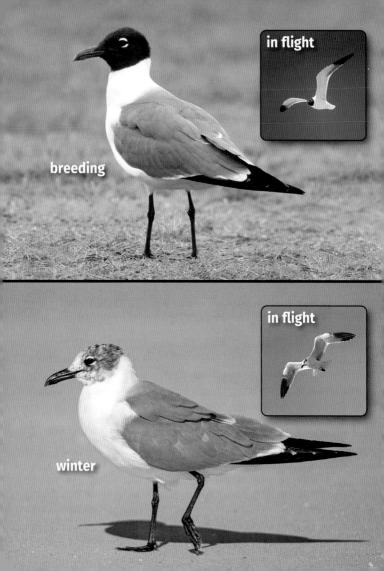

in flight

breeding

in flight

winter

Laughing Gull
Leucophaeus atricilla

YEAR-ROUND WINTER

Size: 16–17" (40–43 cm); up to 3⅓' wingspan

Male: Breeding adult has a black head "hood" and white neck, chest and belly. Slate-gray back and wings with black wing tips. Orange bill. Incomplete white eye-ring. Winter plumage lacks the "hood" and has a black bill.

Female: same as male

Juvenile: brown throughout, gray sides, lacking the black head and white chest, has a gray bill

Nest: ground; male and female construct; 1 brood per year

Eggs: 2–4; olive with brown markings

Incubation: 18–20 days; female and male incubate

Fledging: 30–35 days; male and female feed young

Migration: non-migrator in southern parts of Louisiana and Mississippi; moves inland during winter

Food: fish, insects, aquatic insects

Compare: Ring-billed Gull (p. 343) and Herring Gull (p. 351) are larger. Look for the black head "hood" and slate-gray back and wings of the Laughing Gull.

Stan's Notes: This is a three-year gull that starts out mostly brown and gray. The second year it resembles adults but lacks a complete black head "hood." Breeding plumage in the third year. Male tosses its head back and calls to attract a mate. Nests in marshes in large colonies. Nest is a scrape on the ground lined with grass, sticks and rocks. Adults feed young half-digested food. Name comes from its laughing-like call.

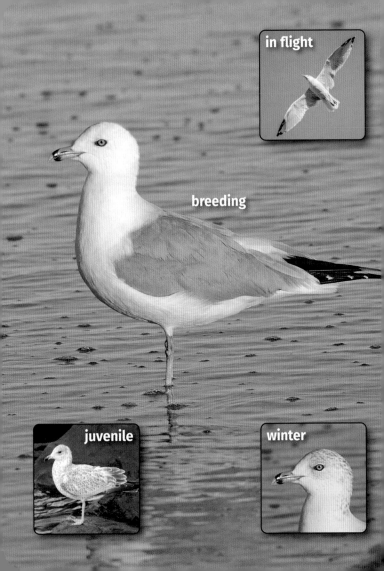

in flight

breeding

juvenile

winter

WINTER

Ring-billed Gull
Larus delawarensis

Size: 18–20" (45–51 cm); up to 4' wingspan

Male: White with gray wings, black wing tips spotted with white, and a white tail, seen in flight (inset). Yellow bill with a black ring near the tip. Yellowish legs and feet. In winter, the back of the head and the nape of the neck are speckled brown.

Female: same as male

Juvenile: white with brown speckles and a brown tip of tail; mostly dark bill

Nest: ground; female and male construct; 1 brood

Eggs: 2–4; off-white with brown markings

Incubation: 20–21 days; female and male incubate

Fledging: 20–40 days; female and male feed the young

Migration: complete, to Louisiana, Mississipi, other southern coastal states, Mexico

Food: insects, fish; scavenges for food

Compare: Laughing Gull (p. 341) has a black head "hood." The Herring Gull (p. 351) has an orange-red mark on its lower bill and pink legs.

Stan's Notes: A common gull of garbage dumps and parking lots and a winter gull in Louisiana and Mississippi. This bird is expanding its range and remaining farther north longer during winter due to successful scavenging in cities. One of the most common gulls in the U.S. A three-year gull with different plumages in each of its first three years. Attains the ring on its bill after the first winter and adult plumage in the third year. Defends a small area around the nest, usually only a few feet.

winter

in flight

breeding

Royal Tern
Sterna maxima

Size: 20" (50 cm)

Male: Gray back and upper surface of wings with white below. Large orange-red bill. Forked tail. Black legs and feet. Breeding plumage has a black cap extending down the nape. Winter plumage has a white forehead and only a partial black cap.

Female: same as male

Juvenile: dull-white to gray with only a hint of a black cap that rarely extends down the nape

Nest: ground; female and male build; 1–2 broods per year

Eggs: 1–2; off-white with dark brown markings

Incubation: 30–31 days; female and male incubate

Fledging: 28–35 days; female and male feed the young

Migration: non-migrator in Louisiana and Mississippi

Food: fish, aquatic insects

Compare: Forster's Tern (p. 339) is smaller and has a small black-tipped bill.

Stan's Notes: A year-round resident in southern Louisiana and Mississippi. Nests in large colonies. Lays one egg (rarely two) in a shallow depression on the ground. Like other terns, Royal Terns plunge from heights 40 feet (12 m) and more into the water headfirst to capture fish and aquatic insects. Populations increase during winter with northern migrants.

in flight

Cattle Egret
Bubulcus ibis

YEAR-ROUND
SUMMER

Size: 18–22" (45–56 cm); up to 3' wingspan

Male: White with orange-buff crest, breast and back. Stocky with a disproportionally large round head. Red-orange bill and legs. Winter plumage is all white with a yellow bill and dark legs.

Female: same as male

Juvenile: similar to winter adult but with a dark bill

Nest: platform; female and male build; 1 brood per year

Eggs: 2–5; light blue-green without markings

Incubation: 22–26 days; female and male incubate

Fledging: 28–30 days; female and male feed the young

Migration: non-migrator to partial in Louisiana and Mississippi; moves to find food

Food: insects, small mammals

Compare: Great Egret (p. 355) is about twice as large and has a much longer neck and a much larger bill. White Ibis (p. 353) has a large down-curved bill.

Stan's Notes: Came to South America from Africa around 1880, reaching Florida in the 1940s. Moved into Louisiana and Mississippi in the mid-1950s. Often seen singularly in pastures, hunting insects at cow and horse pies by wiggling its neck and head back and forth and from side to side, while holding its body still. Then it stabs at prey and tosses it to the back of its mouth. Frequently attracted to field fires to hunt newly exposed animals and insects. In some years it is found as far as northern-tier states and Canada.

in flight

Snowy Egret
Egretta thula

YEAR-ROUND
SUMMER
MIGRATION

Size: 22–26" (56–66 cm); up to 3½' wingspan

Male: All-white bird with black bill. Black legs. Bright-yellow feet. Long feather plumes on head, neck and back during breeding season.

Female: same as male

Juvenile: similar to adult, but backs of legs are yellow

Nest: platform; female and male build; 1 brood per year

Eggs: 3–5; light blue-green without markings

Incubation: 20–24 days; female and male incubate

Fledging: 28–30 days; female and male feed the young

Migration: non-migrator to partial in Louisiana and Mississippi, to southern coastal states and Mexico

Food: aquatic insects, small fish

Compare: Great Egret (p. 355) is much larger and has a yellow bill and black feet. Little Blue Heron (p. 119) is the same size and has a black-tipped gray bill. Look for the black bill and yellow feet of Snowy Egret to help identify.

Stan's Notes: Common in wetlands and often seen with other egrets. Colonies may include up to several hundred nests. Nests are low in shrubs 5–10 feet (1.5–3 m) tall or constructs a nest on the ground, usually mixed among other egret and heron nests. Chicks hatch days apart (asynchronous), leading to starvation of last to hatch. Will actively "hunt" prey by moving around quickly, stirring up small fish and aquatic insects with its feet. In the breeding state, a yellow patch at the base of bill and the yellow feet turn orange-red. Was hunted to near extinction in the late 1800s for its feathers.

in flight

breeding

juvenile

winter

WINTER

Herring Gull
Larus argentatus

Size: 23–26" (58–66 cm); up to 5' wingspan

Male: White with slate-gray wings. Black wing tips with tiny white spots. Yellow bill with an orange-red spot near the tip of the lower bill (mandible). Pinkish legs and feet. Winter plumage has gray speckles on head and neck.

Female: same as male

Juvenile: mottled brown to gray, with a black bill

Nest: ground; female and male construct; 1 brood per year

Eggs: 2–3; olive with brown markings

Incubation: 24–28 days; female and male incubate

Fledging: 35–36 days; female and male feed the young

Migration: complete, to Louisiana, Mississippi, other southern states

Food: fish, insects, clams, eggs, baby birds

Compare: Ring-billed Gull (p. 343) is smaller and has yellowish legs and feet and a black ring on its bill. Look for the orange-red spot on the bill to help identify the Herring Gull.

Stan's Notes: A common gull of large freshwater lakes. An opportunistic bird, scavenging for human food in dumpsters, parking lots and other places with garbage. Takes eggs and young from other bird nests. Often drops clams and other shellfish from heights to break the shells and get to the soft interior. Nests in colonies, returning to the same site annually. Lines its nest with grass and seaweed. It takes about four years for the juveniles to obtain adult plumage. Adults have spotted heads during winter. Adults molt to dirty gray in winter and look similar to juveniles.

in flight

juvenile

White Ibis
Eudocimus albus

YEAR-ROUND
MIGRATION

Size: 23–27" (58–69 cm); up to 3' wingspan

Male: All-white bird with a very long, downward-curving orange-to-red bill. Pink facial skin. Color of legs matches the bill color. Black wing tips, seen only in flight.

Female: same as male, but smaller; downward curve of bill is less than curve of male bill

Juvenile: combination of chocolate-brown and white for the first two years, dull-orange bill

Nest: platform; female and male build; 1 brood

Eggs: 2–3; light blue with dark markings

Incubation: 21–23 days; female and male incubate

Fledging: 28–35 days; female and male feed the young

Migration: complete to non-migrator in Louisiana and Mississippi; moves around to find food

Food: aquatic insects, crustaceans, fish

Compare: Glossy Ibis (p. 235) is brown. Snowy Egret (p. 349) has a straight black bill and bright-yellow feet. Cattle Egret (p. 347) has a short yellow bill. Wood Stork (p. 97) has a bald dark head and thick bill.

Stan's Notes: Increasing in Louisiana and Mississippi over the past 50 years, with inland sightings getting more common. Prefers fresh water over salt water, with crayfish a big part of its diet. White plumage with black wing tips and a bright orange-to-red down-curved bill make this species easy to identify. Frequently seen flying in groups of 30 or more. Nests in large colonies in well-made stick nests. Often seen stalking along just about any wet area in the city or country.

in flight

Great Egret

Ardea alba

YEAR-ROUND
SUMMER

Size: 36–40" (91–102 cm); up to 4½' wingspan

Male: Tall, thin, all-white bird with a long neck and a long, pointed yellow bill. Black, stilt-like legs and black feet.

Female: same as male

Juvenile: same as adults

Nest: platform; male and female construct; 1 brood per year

Eggs: 2–3; light blue without markings

Incubation: 23–26 days; female and male incubate

Fledging: 43–49 days; female and male feed the young

Migration: non-migrator to partial in Louisiana and Mississippi

Food: small fish, aquatic insects, frogs, crayfish

Compare: Cattle Egret (p. 347) is about half the size of Great Egret and has a much shorter neck and much smaller bill. The Snowy Egret (p. 349) is much smaller with yellow feet and a black bill. Juvenile Little Blue Heron (p. 119) is smaller and has a black-tipped gray bill. White Ibis (p. 353) has a very long, down-curved orange-to-red bill.

Stan's Notes: Slowly stalks shallow ponds, lakes and wetlands in search of small fish to spear with its long, sharp bill. The name "Egret" comes from the French word *aigrette,* meaning "ornamental tufts of plumes." The plumes grow near the tail during the breeding season. Hunted to near extinction in the 1800s and early 1900s for its long plumes, which were used to decorate women's hats. Today, the egret is a protected species.

in flight

breeding

chick-feeding adult

American White Pelican
Pelecanus erythrorhynchos

Size: 60–64" (152–163 cm); up to 9' wingspan

Male: Large white pelican with an enormous bright-yellow-to-orange bill. Yellow legs and feet. Black wing tips and trailing edge of wings. Breeding plumage has a bright-orange bill, legs and feet. Chick-feeding adult (an adult that is feeding young) has a gray-black crown.

Female: same as male

Juvenile: duller white than adult, with a brownish head and neck

Nest: ground, scraped-out depression rimmed with dirt; female and male build; 1 brood per year

Eggs: 1–3; white without markings

Incubation: 29–36 days; male and female incubate

Fledging: 60–70 days; female and male feed the young

Migration: complete, to Louisiana and Mississippi

Food: fish

Compare: Brown Pelican (p. 239) has a gray-brown body and a black belly. Look for American White Pelican's black wing tips in flight.

Stan's Notes: Often seen in large groups on the larger freshwater lakes and reservoirs in Louisiana and Mississippi. Doesn't dive to catch fish, like coastal Brown Pelicans. Instead, groups swim and dip their bills simultaneously into water to scoop up fish. Groups fly in a large V, often gliding, followed by simultaneous flapping. Large flocks swirl on columns of rising warm air (thermals) on hot days. Breeding adults typically grow a flat, fibrous plate on the upper bill, which drops off after the eggs hatch.

male

winter male

female

YEAR-ROUND
WINTER

American Goldfinch
Spinus tristis

Size: 5" (13 cm)

Male: Canary-yellow finch with a black forehead and tail. Black wings with white wing bars. White rump. No markings on the chest. Winter male is similar to the female.

Female: dull olive-yellow plumage with brown wings; lacks a black forehead

Juvenile: same as female

Nest: cup; female builds; 1 brood per year

Eggs: 4–6; pale blue without markings

Incubation: 10–12 days; female incubates

Fledging: 11–17 days; female and male feed the young

Migration: partial to non-migrator; small flocks of up to 20 birds move around to find food

Food: seeds, insects; will come to seed feeders

Compare: The Pine Siskin (p. 129) has a streaked chest and belly and yellow wing bars. The female House Finch (p. 131) and female Purple Finch (p. 145) have heavily streaked chests.

Stan's Notes: A common backyard resident. Most often found in open fields, scrubby areas and woodlands. Enjoys Nyjer seed in feeders. Breeds in late summer. Lines its nest with the silky down from wild thistle. Almost always in small flocks. Twitters while it flies. Flight is roller coaster-like. Moves around to find adequate food during winter. Starts moving to Louisiana and Mississippi in November. Can be a common visitor to feeders during winter, leaving in April for northern states. Often called Wild Canary due to the male's canary-colored plumage. Male sings a pleasant, high-pitched song. Very numerous in some years and scarce in others.

male

female

Common Yellowthroat

Geothlypis trichas

YEAR-ROUND
SUMMER

Size: 5" (13 cm)

Male: Olive-brown with a bright-yellow throat and chest, a white belly and a distinctive black mask outlined in white. Long, thin, pointed black bill.

Female: similar to male but lacks a black mask

Juvenile: same as female

Nest: cup; female builds; 2 broods per year

Eggs: 3–5; white with brown markings

Incubation: 11–12 days; female incubates

Fledging: 10–11 days; female and male feed the young

Migration: complete, to Louisiana and Mississippi, other southern states, Mexico and Central America; non-migrator in Louisiana and much of Mississippi

Food: insects

Compare: The male American Goldfinch (p. 359) has a black forehead and wings. The Yellow-rumped Warbler (p. 251) only has patches of yellow and lacks the yellow chest of the Yellowthroat.

Stan's Notes: A common warbler of open fields and marshes. Sings a cheerful, well-known "witchity-witchity-witchity-witchity" song from deep within tall grasses. Male sings from prominent perches and while he hunts. He performs a curious courtship display, bouncing in and out of tall grass while singing a mating song. Female builds a nest low to the ground. Young remain dependent on their parents longer than most other warblers. A frequent cowbird host.

male
p. 321

female

American Redstart
Setophaga ruticilla

SUMMER
MIGRATION

Size: 5" (13 cm)

Female: Olive-brown warbler with yellow patches on the sides, wings and tail. White belly.

Male: black with orange patches on the sides, wings and tail; white belly

Juvenile: same as female; male attains orange tinges in the second year

Nest: cup; female builds; 1 brood per year

Eggs: 3–5; off-white with brown markings

Incubation: 12 days; female incubates

Fledging: 9 days; female and male feed the young

Migration: complete, to Mexico, Central America and South America

Food: insects, seeds, occasionally berries

Compare: The female Yellow-rumped Warbler (p. 251) is similar, but it has a yellow patch on its rump. Look for yellow patches on the sides, wings and tail to help identify the female Redstart.

Stan's Notes: A common widespread warbler in Louisiana and Mississippi during summer and migration. Prefers large, unbroken tracts of forest. Found in woodlands, parks and yards and at forest edges. Appears hyperactive when it feeds, hovering and darting back and forth to glean insects from leaves. Often droops wings and fans tail before launching out to catch an insect. Look for the flashing black-and-orange colors of the male high up in trees.

Prairie Warbler
Setophaga discolor

Size: 5" (13 cm)

Male: Olive back with chestnut-colored streaks. Bright yellow from the chin to belly. Black streaks on sides from neck down. Black line through eyes. Yellow eyebrows.

Female: same as male, only duller

Juvenile: similar to female

Nest: cup; female builds; 2 broods per year

Eggs: 3–5; white with brown markings

Incubation: 11–14 days; female incubates

Fledging: 8–11 days; female and male feed young

Migration: complete, to Florida and the Caribbean

Food: insects

Compare: Male Common Yellowthroat (p. 361) has a complete black mask. Watch for Prairie Warbler to twitch its tail when feeding.

Stan's Notes: A common resident and migrator in Louisiana and Mississippi. Returns to Louisiana and Mississippi in mixed flocks of warblers in late April to mid-May. It was unfortunately misnamed "Prairie" when first found in a barren area in Kentucky. Nests in dry, brushy clearings and forest edges, making it a perfect host for Brown-headed Cowbirds. Will sometimes desert a parasitized nest. Nests in upright fork of a tree. Feeds young mainly caterpillars.

winter

MIGRATION
WINTER

Palm Warbler
Setophaga palmarum

Size: 5½" (14 cm)

Male: Distinctive yellow eyebrows. Yellow throat, belly and undertail. Obvious chestnut cap. Thin chestnut streaks on the sides of the breast. Dark line across dark eyes. Winter plumage similar but duller, often lacking brown cap.

Female: same as male

Juvenile: same as adult but duller and brown

Nest: cup; female builds; 1–2 broods per year

Eggs: 4–5; white with brown markings

Incubation: 11–12 days; female incubates

Fledging: 12–13 days; female and male feed the young

Migration: complete, to Louisiana, Mississippi, other southern states, the Caribbean and Central America

Food: insects, fruit

Compare: The Yellow-rumped Warbler (p. 251) is similar in size but lacks the yellow throat and belly of the Palm. The Pine Warbler (p. 369) has pronounced white wing bars. Look for the yellow eyebrows and chestnut cap of the Palm Warbler.

Stan's Notes: A common bird in Louisiana and Mississippi during migration and winter. Look for it to wag or bob its tail while gleaning insects from leaves and flowers of trees. One of the few warblers to feed on the ground. Recognizes cowbird eggs, which it either rejects from the nest and destroys or buries under a new nest built on top of the old nest and eggs.

Pine Warbler
Setophaga pinus

YEAR-ROUND
SUMMER
WINTER

Size: 5½" (14 cm)

Male: A yellow throat and breast with faint black streaks on sides of breast. Olive-green back. Two white wing bars. White belly.

Female: similar to male, only paler

Juvenile: similar to adults, but is browner with more white on belly

Nest: cup; female builds; 2–3 broods per year

Eggs: 3–5; white with brown markings

Incubation: 10–12 days; female incubates

Fledging: 12–14 days; female and male feed the young

Migration: partial to non-migrator in Louisiana and Mississippi; moves around during winter to find food

Food: insects, seeds, fruit

Compare: Palm Warbler (p. 367) is similar, but it has a brown cap and yellow eyebrows. Pine Warbler has much more pronounced white wing bars than the Palm Warbler. The Yellow-rumped Warbler (p. 251) has yellow patches on its rump. The American Goldfinch (p. 359) lacks streaks on breast.

Stan's Notes: A common resident of pine forests in Louisiana and Mississippi. Nests only in pine forest. Brighter in spring and more drab in fall, it varies in color depending upon the time of year. Thought to have a larger bill than the other warblers. One of the few warbler species found year round in some parts of both states. Populations increase each fall and winter with the arrival of northern birds.

male
p. 323

female

Baltimore Oriole
Icterus galbula

SUMMER MIGRATION

Size: 7–8" (18–20 cm)

Female: Pale yellow with orange tones and gray-brown wings with white wing bars. Gray bill. Dark eyes.

Male: flaming orange with a black head and back, white-and-orange wing bars, an orange-and-black tail, a gray bill and dark eyes

Juvenile: same as female

Nest: pendulous; female builds; 1 brood per year

Eggs: 4–5; bluish with brown markings

Incubation: 12–14 days; female incubates

Fledging: 12–14 days; female and male feed the young

Migration: complete, to Mexico, Central America and South America

Food: insects, fruit, nectar; comes to nectar, orange-half, and grape-jelly feeders

Compare: The female Orchard Oriole (p. 373) has a dull-yellow belly. Look for the gray-brown wings to identify the female Baltimore Oriole.

Stan's Notes: A fantastic songster, often heard before seen. Easily attracted to bird feeders that offer sugar water (nectar), orange halves or grape jelly. Parents bring young to feeders. Sits at the tops of trees, feeding on caterpillars. Female builds a sock-like nest at the outermost branches of tall trees. Prefers parks, yards and forests and often returns to the same area year after year. Some of the last birds to arrive in spring and first to leave in fall. Seen during migration and summer.

male
p. 325

female

first-year
male

Orchard Oriole

Icterus spurius

SUMMER

Size: 7–8" (18–20 cm)

Female: Olive-green with a dull-yellow belly. Gray wings with 2 indistinct white wing bars. Long, thin bill with a gray mark on the lower bill.

Male: dark orange with black head, throat, upper back, wings and tail; 1 white wing bar

Juvenile: same as female; first-year male looks like the female, with a black bib

Nest: pendulous; female builds; 1 brood per year

Eggs: 3–5; pale blue to white, brown markings

Incubation: 11–12 days; female and male incubate

Fledging: 11–14 days; female and male feed the young

Migration: complete, to Mexico, Central America and northern South America

Food: insects, fruit, nectar; comes to nectar, orange-half and grape-jelly feeders

Compare: Female Baltimore Oriole (p. 371) is similar but has orange tones and more-distinct wing bars. The female Summer Tanager (p. 375) is mustard-yellow with a larger bill.

Stan's Notes: Named "Orchard" for its preference for orchards. Also likes open woods. Eats insects until wild fruit starts to ripen. Summer resident, arriving in early spring and departing at the end of summer. Often nests alone; sometimes nests in small colonies. Parents bring their young to bird feeding stations after they fledge. Many people don't see these birds at feeders much during the summer and think they have left, but the birds are still there, hunting for insects to feed to their young. Often migrates with Baltimore Orioles.

male
p. 331

female

SUMMER

Summer Tanager
Piranga rubra

Size: 8" (20 cm)

Female: Some show a faint wash of red, but most females are a mustard-yellow overall with slightly darker wings.

Male: bright rosy-red bird with darker red wings

Juvenile: male has patches of red and green over the entire body, female is same as adult female

Nest: cup; female builds; 1–2 broods per year

Eggs: 3–5; pale blue with dark markings

Incubation: 10–12 days; female incubates

Fledging: 12–15 days; female and male feed young

Migration: complete, to Mexico and Central and South America

Food: insects, fruit

Compare: Similar to female Orchard Oriole (p. 373) and Baltimore Oriole (p. 371). Look for Summer Tanager's lack of wing bars and larger, thicker bill to identify.

Stan's Notes: Found throughout Louisiana and Mississippi where woodlands exist, especially in mixed pine and oak forests. Due to clearing of land for agriculture, populations have decreased for over a century and especially most recently. Returning to Louisiana and Mississippi in late March and with young hatching in late May, some pairs have two broods per year. Most leave by November with very few staying for winter. While fruit makes up some of the diet, most of it consists of insects such as bees and wasps. Summer Tanagers unfortunately seem to be parasitized by Brown-headed Cowbirds more than any other nesting bird in Louisiana and Mississippi.

YEAR-ROUND

Eastern Meadowlark
Sturnella magna

Size: 9" (23 cm)

Male: Robin-shaped bird with a brown back and yellow chest and belly. V-shaped black necklace. Short tail with white outer tail feathers, best seen when flying away.

Female: same as male

Juvenile: same as adult

Nest: cup, on the ground in dense cover; female builds; 2 broods per year

Eggs: 3–5; white with brown markings

Incubation: 13–15 days; female incubates

Fledging: 11–12 days; female and male feed the young

Migration: non-migrator in Louisiana and Mississippi

Food: insects, seeds

Compare: This is the only large yellow bird that has a black V mark on the breast.

Stan's Notes: A songbird of open grassy country, singing when perched and in flight. Given the name "Meadowlark" because it's a bird of meadows and sings like the larks of Europe. Best known for its wonderful, clear, flute-like whistling song. Often seen perching on fence posts but will quickly dive into tall grass when approached. Conspicuous white markings on each side of its tail, most often seen when flying away. Sometimes domes its nest with dried grass. Not in the lark family. A member of the blackbird family, related to grackles and orioles. Populations increase in winter when migratory birds from the north join resident birds. Overall population is down greatly as a result of intensive agricultural activities and ditch mowing.

BIRDING ON THE INTERNET

Birding online is a great way to discover additional information and learn more about birds. These websites will assist you in your pursuit of birds. Web addresses sometimes change a bit, so if one no longer works, just enter the name of the group into a search engine to track down the new address.

Site	Address
Author Stan Tekiela's homepage	naturesmart.com
American Birding Association	aba.org
Louisiana Audubon	la.audubon.org
Mississippi Audubon	ms.audubon.org
Louisiana Ornithological Society	www.losbird.org
Mississippi Ornithological Society	www.missbird.org
The Cornell Lab of Ornithology	birds.cornell.edu
ebird	ebird.org
Acadiana Wildlife Education & Rehabilitation, Inc.	acadianawildlife.org
Mississippi Wildlife Rehabilitation, Inc.	www.mswildliferehab.org
The Jackson Zoo Raptor Rehab	jacksonzoo.org/conservation /raptor-rehab-2/

CHECKLIST/INDEX BY SPECIES

Use the boxes to check the birds you've seen.

MORE FOR LOUISIANA AND MISSISSIPPI BY STAN TEKIELA

Identification Guides
Birds of Prey of the South Field Guide

Birds of the South

Shorebirds of the Southeast & Gulf States

Stan Tekiela's Birding for Beginners: South

Children's Books: Adventure Board Book Series
Floppers & Loppers

Paws & Claws

Peepers & Peekers

Snouts & Sniffers

Children's Books
C is for Cardinal

Can You Count the Critters?

Critter Litter

Children's Books: Wildlife Picture Books
Baby Bear Discovers the World

The Cutest Critter

Do Beavers Need Blankets?

Hidden Critters

Jump, Little Wood Ducks

Some Babies Are Wild

Super Animal Powers

What Eats That?

Whose Baby Butt?

Whose Butt?

Whose House Is That?

Whose Track Is That?

Our Love of Wildlife Series
Our Love of Loons

Our Love of Owls

Nature Books
Bird Trivia
Start Mushrooming
A Year in Nature with Stan Tekiela

Favorite Wildlife Series
Bald Eagles
Hummingbirds
Loons

Wildlife Appreciation Series
Backyard Birds
Bears
Bird Migration
Cranes, Herons & Egrets
Deer, Elk & Moose
Wild Birds

Nature Appreciation Series
Bird Nests
Feathers
Wildflowers

Nature's Wild Cards (playing cards)
Bears
Birds of the Gulf Coast
Hummingbirds
Loons
Mammals of the Gulf Coast
Owls
Raptors
Trees of the Gulf Coast

ABOUT THE AUTHOR

Naturalist, wildlife photographer and writer Stan Tekiela is the originator of the popular state-specific field guide series that includes the *Birds of Alabama Field Guide*. Stan has authored more than 190 educational books, including field guides, quick guides, nature books, children's books and more, presenting many species of animals and plants.

With a Bachelor of Science degree in natural history from the University of Minnesota and as an active professional naturalist for more than 30 years, Stan studies and photographs wildlife throughout the United States and Canada. He has received national and regional awards for his books and photographs and is also a well-known columnist and radio personality. His syndicated column appears in more than 25 newspapers, and his wildlife programs are broadcast on a number of Midwest radio stations. You can follow Stan on Facebook and Twitter or contact him via his website, naturesmart.com.